2018: An Astronomical Year

A Reference Guide to 365 Nights of Astronomy

Richard J. Bartlett

Cover Image: A long series of exposures at night captures the light trails left by the moon, planets and the stars through the night sky.

Credit: ESO/B. Tafreshi (twanight.org)

Image source: http://www.eso.org/public/images/trailconjunction-cc/

Planet images created using the *Mobile Observatory* app for Android devices. http://zima.co

Contents

About the 2018 Edition

Last year saw a major change in the format of the guide and I've made a few more changes for this edition. One big change last year was the addition of data tables for the Moon and planets and depictions of the view through a telescope. These provided a good overview, such as their magnitude, phase, apparent size etc, but I wasn't able to properly illustrate where the planets were in the sky. As a result, this year I've included charts to show the location of Mercury, Venus, Mars, Jupiter, Saturn and Uranus for each of the ten day periods in each month. That way, besides knowing that Venus is close to Aldebaran and the Hyades, you'll also be able to see exactly where it is in relation to those objects.

Another big change is that I've streamlined the events described in the text. Previously, for the sake of completeness, I'd included events that weren't always visible or of minor importance. For this edition I've stuck to listing the major events (such as solar conjunctions) as well as the events you can see with your eyes or, in some circumstances, binoculars or a telescope.

The summary tables have also changed format a little. Hopefully they're a little easier to understand but the principle still applies. The table shows the relevant phases of the Moon for each ten day period and also shows which planets appear close to each other or the Moon during that time.

Lastly, in another big change, I've removed the star charts for each ten day period and replaced them with 24 charts you can use (theoretically) at any time of year. I did this because the charts in the previous edition could only be used at a specific time of night and at a specific time of year – for example, 10pm on March 15th. With these charts you can get an idea of which constellations are visible between 6pm and 6am throughout the entire year. I've also included a list of some fifty multiple stars, carbon(red)/variable stars and deep sky objects that should also be visible at that time. For the record, it's by no means a complete list as it only lists the brightest objects but it should provide a good starting point for a night's observing.

The downside is that the charts don't have the planets indicated but, again, there are the star charts with the planetary data to help observers with that.

As always, the data and charts were compiled using Wolfgang Zima's excellent *Mobile Observatory* app for Android devices. You can download it from the Google Play store or check it out at http://zima.co

The images of the planets were also created using the app. Each image shows the view you might get through a very high magnification eyepiece and spans the equivalent of one arc second. That way you can get an idea of how much larger or smaller a planet might appear in comparison to another.

I'm always looking for ways to improve the book and am open to comments and suggestions. With that in mind, please feel free to email me at astronomywriter@gmail.com if there's anything you'd like to see included.

About the Author

Photo by my son, James Bartlett

I've had an interest in astronomy since I was six and although my interest has waxed and waned like the Moon, I've always felt compelled to stop and stare at the stars.

In the late 90's, I discovered the booming frontier of the internet, and like a settler in the Midwest, I quickly staked my claim on it. I started to build a (now-defunct) website called *StarLore.* It was designed to be an online resource for amateur astronomers who wanted to know more about the constellations - and all the stars and deep sky objects to be found within them. It was quite an undertaking.

After the website was featured in the February 2001 edition of *Sky & Telescope* magazine, I began reviewing astronomical websites and software for their rival, *Astronomy.* This was something of a dream come true; I'd been reading the magazine since I was a kid and now my name was regularly appearing in it.

Unfortunately, a financial downturn forced my monthly column to be cut after a few years but I'll always be grateful for the chance to write for the world's best-selling astronomy magazine.

I emigrated from England to the United States in 2004 and spent three years under relatively clear, dark skies in Oklahoma. I then relocated to Kentucky in 2008 and then California in 2013. I now live in the suburbs of Los Angeles; not the most ideal location for astronomy, but there are still a number of naked eye events that are easily visible on any given night.

The Author Online

Amazon US: http://tinyurl.com/rjbamazon-us

Amazon UK: http://tinyurl.com/rjbamazon-uk

Facebook: http://tinyurl.com/rjbfacebook

Twitter: http://tinyurl.com/rjbtwitter

Blog: http://tinyurl.com/astronomicalyear

Email: astronomywriter@gmail.com

AstroNews: http://tinyurl.com/astronewsus

Clear skies,

Richard J. Bartlett

May 24th, 2017

Star Chart Tables

If observing during daylight savings time, first deduct one hour and then refer to the corresponding chart number. For example, for 10pm daylight savings time in early August, use chart 18.

	6pm	7pm	8pm	9pm	10pm	11pm
Early January	1	2	3	4	5	6
Late January	2	3	4	5	6	7
Early February	3	4	5	6	7	8
Late February	4	5	6	7	8	9
Early March	5	6	7	8	9	10
Late March	6	7	8	9	10	11
Early April	7	8	9	10	11	12
Late April	8	9	10	11	12	13
Early May	9	10	11	12	13	14
Late May	10	11	12	13	14	15
Early June	11	12	13	14	15	16
Late June	12	13	14	15	16	17
Early July	13	14	15	16	17	18
Late July	14	15	16	17	18	19
Early August	15	16	17	18	19	20
Late August	16	17	18	19	20	21
Early September	17	18	19	20	21	22
Late September	18	19	20	21	22	23
Early October	19	20	21	22	23	24
Late October	20	21	22	23	24	1
Early November	21	22	23	24	1	2
Late November	22	23	24	1	2	3
Early December	23	24	1	2	3	4
Late December	24	1	2	3	4	5

If observing during daylight savings time, first deduct one hour and then refer to the corresponding chart number. For example, for 2am daylight savings time in early July, use chart 20.

	12am	1am	2am	3am	4am	5am	6am
Early January	7	8	9	10	11	12	13
Late January	8	9	10	11	12	13	14
Early February	9	10	11	12	13	14	15
Late February	10	11	12	13	14	15	16
Early March	11	12	13	14	15	16	17
Late March	12	13	14	15	16	17	18
Early April	13	14	15	16	17	18	19
Late April	14	15	16	17	18	19	20
Early May	15	16	17	18	19	20	21
Late May	16	17	18	19	20	21	22
Early June	17	18	19	20	21	22	23
Late June	18	19	20	21	22	23	24
Early July	19	20	21	22	23	24	1
Late July	20	21	22	23	24	1	2
Early August	21	22	23	24	1	2	3
Late August	22	23	24	1	2	3	4
Early September	23	24	1	2	3	4	5
Late September	24	1	2	3	4	5	6
Early October	1	2	3	4	5	6	7
Late October	2	3	4	5	6	7	8
Early November	3	4	5	6	7	8	9
Late November	4	5	6	7	8	9	10
Early December	5	6	7	8	9	10	11
Late December	6	7	8	9	10	11	12

Chart 1

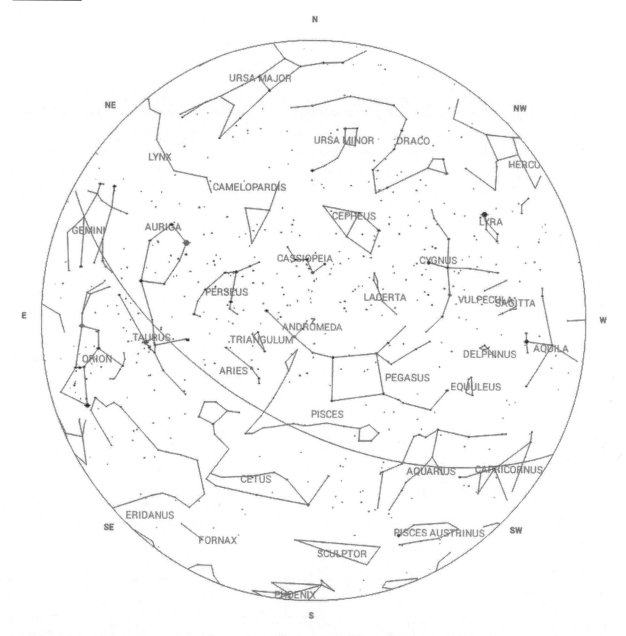

Designation	Name	Con.	Type	R.A.	Dec.	Mag	Size/Sep
Gam And	Almach	And	MS	02h 04m	+42° 20'	2.1	10"
M 31	Andromeda Galaxy	And	Gx	00h 43m	+41° 16'	4.3	156'
NGC 752	Golf Ball Cluster	And	OC	01h 58m	+37° 47'	6.6	75'
Pi And		And	MS	00h 38m	+33° 49'	4.4	36"
107 Aqr		Aqr	MS	23h 47m	-18° 35'	5.3	7"
94 Aqr		Aqr	MS	23h 19m	-13° 28'	5.2	13"
NGC 7293	Helix Nebula	Aqr	PN	22h 30m	-20° 50'	6.3	16'
Zet Aqr		Aqr	MS	22h 29m	-00° 01'	3.7	2"

Designation	Name	Con.	Type	R.A.	Dec.	Mag	Size/Sep
41 Aqr		Aqr	MS	22h 14m	-21° 04'	5.3	5"
53 Aqr		Aqr	MS	22h 27m	-16° 45'	5.6	3"
Lam Ari		Ari	MS	01h 59m	+23° 41'	4.8	37"
Gam Ari	Mesarthim	Ari	MS	01h 54m	+19° 22'	4.6	8"
30 Ari		Ari	MS	02h 37m	+24° 39'	6.5	39"
Kemble 1	Kemble's Cascade	Cam	Ast	03h 57m	+63° 04'	5.0	180'
Sig Cas		Cas	MS	23h 59m	+55° 45'	4.9	3"
NGC 457	Owl Cluster	Cas	OC	01h 20m	+58° 17'	5.1	20'
M 52		Cas	OC	23h 25m	+61° 36'	8.2	15'
Struve 163		Cas	MS	01h 51m	+64° 51'	6.5	35"
Struve 3053		Cas	MS	00h 03m	+66° 06'	5.9	15"
Eta Cas	Achird	Cas	MS	00h 50m	+57° 54'	3.6	13"
M 103		Cas	OC	01h 33m	+60° 39'	6.9	5'
NGC 281		Cas	OC	00h 53m	+56° 38'	7.4	4'
Iot Cas		Cas	MS	02h 29m	+67° 24'	4.5	7"
NGC 7789	Herschel's Spiral Cluster	Cas	OC	23h 57m	+56° 43'	7.5	25'
NGC 559		Cas	OC	01h 30m	+63° 18'	7.4	6'
NGC 659	Ying Yang Cluster	Cas	OC	01h 44m	+60° 40'	7.2	5'
NGC 663		Cas	OC	01h 46m	+61° 14'	6.4	14'
Del Cep		Cep	MS/Var	22h 29m	+58° 25'	3.5-4.4	41"
Xi Cep	Alkurhah	Cep	MS	22h 04m	+64° 38'	4.3	8"
Omi Cep		Cep	MS	23h 19m	+68° 07'	4.8	3"
Gam Cet	Kaffajidhma	Cet	MS	02h 43m	+03° 14'	3.5	3"
Omi Cet	Mira	Cet	Var	02h 19m	-02° 59'	2.0-10.1	N/A
NGC 7243		Lac	OC	22h 15m	+49° 54'	6.7	29'
8 Lac		Lac	MS	22h 36m	+39° 38'	5.7	82"
Bet Per	Algol	Per	Var	03h 08m	+40° 57'	2.1-3.4	N/A
Eps Per		Per	MS	03h 58m	+40° 01'	2.9	9"
M 34		Per	OC	02h 42m	+42° 46'	5.8	35'
NGC 869/884	Double Cluster	Per	OC	02h 21m	+57° 08'	4.4	18'
Eta Per		Per	MS	02h 51m	+55° 54'	3.8	29"
NGC 1245		Per	OC	03h 15m	+47° 15'	7.7	10'
Melotte 20	Alpha Persei Moving Cluster	Per	OC	03h 24m	+49° 52'	2.3	300'
TX Psc		Psc	Var/CS	23h 46m	+03° 29'	4.5-5.3	N/A
Alp Psc	Alrisha	Psc	MS	02h 02m	+02° 46'	3.8	2"
Psi1 Psc		Psc	MS	01h 06m	+21° 28'	5.3	30"
Zet Psc		Psc	MS	01h 14m	+07° 35'	5.2	23"
55 Psc		Psc	MS	00h 40m	+21° 26'	5.4	6"
65 Psc		Psc	MS	00h 50m	+27° 43'	7.0	4"
M 45	Pleiades	Tau	OC	03h 47m	+24° 07'	1.5	120'
M 33	Triangulum Galaxy	Tri	Gx	01h 34m	+30° 40'	6.4	62'
Alp UMi	Polaris	UMi	MS	02h 51m	+89° 20'	2.0	18"

Chart 2

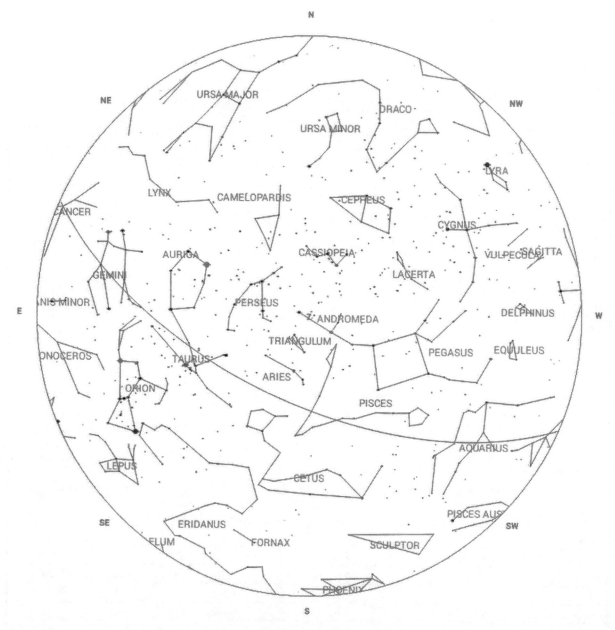

Designation	Name	Con.	Type	R.A.	Dec.	Mag	Size/Sep
Gam And	Almach	And	MS	02h 04m	+42° 20'	2.1	10"
M 31	Andromeda Galaxy	And	Gx	00h 43m	+41° 16'	4.3	156'
NGC 752	Golf Ball Cluster	And	OC	01h 58m	+37° 47'	6.6	75'
NGC 7662	Blue Snowball Nebula	And	PN	23h 26m	+42° 32'	8.6	17"
Pi And		And	MS	00h 38m	+33° 49'	4.4	36"
Lam Ari		Ari	MS	01h 59m	+23° 41'	4.8	37"
Gam Ari	Mesarthim	Ari	MS	01h 54m	+19° 22'	4.6	8"
30 Ari		Ari	MS	02h 37m	+24° 39'	6.5	39"

Designation	Name	Con.	Type	R.A.	Dec.	Mag	Size/Sep
U Cam		Cam	Var/CS	03h 42m	+62° 39'	7.0-7.5	N/A
ST Cam		Cam	Var/CS	04h 51m	+68° 10'	7.0-8.4	N/A
Kemble 1	Kemble's Cascade	Cam	Ast	03h 57m	+63° 04'	5.0	180'
NGC 1502	Jolly Roger Cluster	Cam	OC	04h 08m	+62° 20'	4.1	8'
NGC 457	Owl Cluster	Cas	OC	01h 20m	+58° 17'	5.1	20'
Iot Cas		Cas	MS	02h 29m	+67° 24'	4.5	7"
Sig Cas		Cas	MS	23h 59m	+55° 45'	4.9	3"
Struve 163		Cas	MS	01h 51m	+64° 51'	6.5	35"
Struve 3053		Cas	MS	00h 03m	+66° 06'	5.9	15"
Eta Cas	Achird	Cas	MS	00h 50m	+57° 54'	3.6	13"
M 103		Cas	OC	01h 33m	+60° 39'	6.9	5'
M 52		Cas	OC	23h 25m	+61° 36'	8.2	15'
NGC 281		Cas	OC	00h 53m	+56° 38'	7.4	4'
NGC 559		Cas	OC	01h 30m	+63° 18'	7.4	6'
NGC 659	Ying Yang Cluster	Cas	OC	01h 44m	+60° 40'	7.2	5'
NGC 663		Cas	OC	01h 46m	+61° 14'	6.4	14'
NGC 7789	Herschel's Spiral Cluster	Cas	OC	23h 57m	+56° 43'	7.5	25'
Omi Cep		Cep	MS	23h 19m	+68° 07'	4.8	3"
Gam Cet	Kaffajidhma	Cet	MS	02h 43m	+03° 14'	3.5	3"
Omi Cet	Mira	Cet	Var	02h 19m	-02° 59'	2.0-10.1	N/A
32 Eri		Eri	MS	03h 54m	-02° 57'	4.4	7"
40 Eri	Keid	Eri	MS	04h 15m	-07° 39'	4.4	83"
NGC 1528	m & m Double Cluster	Per	OC	04h 15m	+51° 13'	6.4	16'
Bet Per	Algol	Per	Var	03h 08m	+40° 57'	2.1-3.4	N/A
Eps Per		Per	MS	03h 58m	+40° 01'	2.9	9"
NGC 1245		Per	OC	03h 15m	+47° 15'	7.7	10'
M 34		Per	OC	02h 42m	+42° 46'	5.8	35'
NGC 869/884	Double Cluster	Per	OC	02h 21m	+57° 08'	4.4	18'
Eta Per		Per	MS	02h 51m	+55° 54'	3.8	29"
Melotte 20	Alpha Persei Moving Cluster	Per	OC	03h 24m	+49° 52'	2.3	300'
Alp Psc	Alrisha	Psc	MS	02h 02m	+02° 46'	3.8	2"
Psi1 Psc		Psc	MS	01h 06m	+21° 28'	5.3	30"
Zet Psc		Psc	MS	01h 14m	+07° 35'	5.2	23"
55 Psc		Psc	MS	00h 40m	+21° 26'	5.4	6"
65 Psc		Psc	MS	00h 50m	+27° 43'	7.0	4"
TX Psc		Psc	Var/CS	23h 46m	+03° 29'	4.5-5.3	N/A
M 45	Pleiades	Tau	OC	03h 47m	+24° 07'	1.5	120'
Melotte 25	Hyades	Tau	OC	04h 27m	+15° 52'	0.8	330'
NGC 1647	Pirate Moon Cluster	Tau	OC	04h 46m	+19° 07'	6.2	40'
M 33	Triangulum Galaxy	Tri	Gx	01h 34m	+30° 40'	6.4	62'
Alp UMi	Polaris	UMi	MS	02h 51m	+89° 20'	2.0	18"

Chart 3

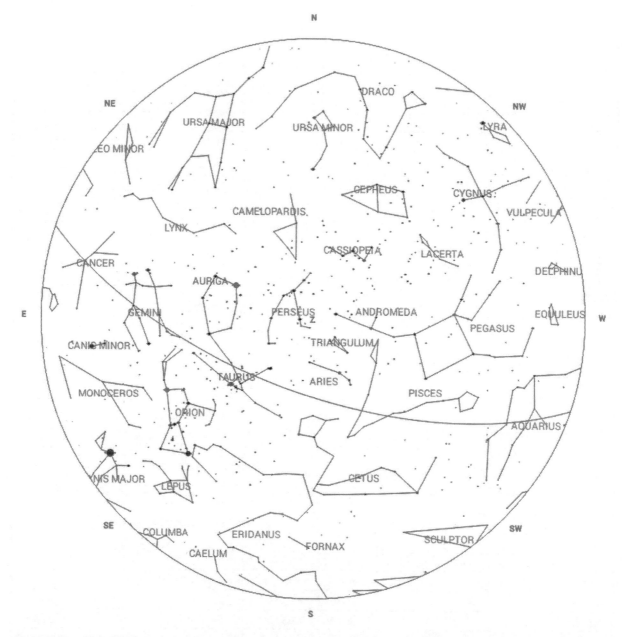

Designation	Name	Con.	Type	R.A.	Dec.	Mag	Size/Sep
Gam And	Almach	And	MS	02h 04m	+42° 20'	2.1	10"
M 31	Andromeda Galaxy	And	Gx	00h 43m	+41° 16'	4.3	156'
Pi And		And	MS	00h 38m	+33° 49'	4.4	36"
Lam Ari		Ari	MS	01h 59m	+23° 41'	4.8	37"
Gam Ari	Mesarthim	Ari	MS	01h 54m	+19° 22'	4.6	8"
30 Ari		Ari	MS	02h 37m	+24° 39'	6.5	39"
M 37		Aur	OC	05h 52m	+32° 33'	6.2	14'
M 36		Aur	OC	05h 36m	+34° 08'	6.5	10'

Designation	Name	Con.	Type	R.A.	Dec.	Mag	Size/Sep
14 Aur		Aur	MS	05h 15m	+32° 41'	5.0	15"
Kemble 1	Kemble's Cascade	Cam	Ast	03h 57m	+63° 04'	5.0	180'
NGC 1502	Jolly Roger Cluster	Cam	OC	04h 08m	+62° 20'	4.1	8'
NGC 457	Owl Cluster	Cas	OC	01h 20m	+58° 17'	5.1	20'
Iot Cas		Cas	MS	02h 29m	+67° 24'	4.5	7"
Struve 3053		Cas	MS	00h 03m	+66° 06'	5.9	15"
Eta Cas	Achird	Cas	MS	00h 50m	+57° 54'	3.6	13"
NGC 663		Cas	OC	01h 46m	+61° 14'	6.4	14'
Gam Cet	Kaffajidhma	Cet	MS	02h 43m	+03° 14'	3.5	3"
Omi Cet	Mira	Cet	Var	02h 19m	-02° 59'	2.0-10.1	N/A
32 Eri		Eri	MS	03h 54m	-02° 57'	4.4	7"
40 Eri	Keid	Eri	MS	04h 15m	-07° 39'	4.4	83"
Sig Ori		Ori	MS	05h 40m	-02° 36'	3.8	42"
Bet Ori	Rigel	Ori	MS	05h 14m	-08° 12'	0.2	10"
Alp Ori	Betelgeuse	Ori	Var	05h 55m	+07° 24'	0.4-1.3	N/A
Eta Ori		Ori	MS	05h 25m	-02° 24'	3.3	2"
Zet Ori	Alnitak	Ori	MS	05h 41m	-01° 57'	1.8	3"
23 Ori		Ori	MS	05h 23m	+03° 33'	5.0	32"
Iot Ori	Nair al Saif	Ori	MS	05h 35m	-05° 55'	2.8	11"
Collinder 70	Epsilon Orionis Cluster	Ori	OC	05h 36m	-01° 00'	0.4	150'
Struve 747		Ori	MS	05h 35m	-05° 55'	4.8	36"
Collinder 72		Ori	OC	05h 35m	-05° 55'	2.5	20'
Del Ori	Mintaka	Ori	MS	05h 33m	-00° 18'	2.1	53"
Lam Ori	Meissa	Ori	MC	05h 36m	+09° 56'	3.4	4"
M 42	Orion Nebula	Ori	Neb	05h 35m	-05° 23'	4.0	40'
NGC 1981	Coal Car Cluster	Ori	OC	05h 35m	-04° 26'	4.2	28'
Collinder 69	Lambda Orionis Cluster	Ori	OC	05h 35m	+09° 56'	2.8	70'
NGC 1528	m & m Double Cluster	Per	OC	04h 15m	+51° 13'	6.4	16'
Bet Per	Algol	Per	Var	03h 08m	+40° 57'	2.1-3.4	N/A
Eps Per		Per	MS	03h 58m	+40° 01'	2.9	9"
M 34		Per	OC	02h 42m	+42° 46'	5.8	35'
NGC 869/884	Double Cluster	Per	OC	02h 21m	+57° 08'	4.4	18'
Eta Per		Per	MS	02h 51m	+55° 54'	3.8	29"
Melotte 20	Alpha Persei Moving Cluster	Per	OC	03h 24m	+49° 52'	2.3	300'
Alp Psc	Alrisha	Psc	MS	02h 02m	+02° 46'	3.8	2"
Psi1 Psc		Psc	MS	01h 06m	+21° 28'	5.3	30"
Zet Psc		Psc	MS	01h 14m	+07° 35'	5.2	23"
55 Psc		Psc	MS	00h 40m	+21° 26'	5.4	6"
M 45	Pleiades	Tau	OC	03h 47m	+24° 07'	1.5	120'
Melotte 25	Hyades	Tau	OC	04h 27m	+15° 52'	0.8	330'
NGC 1647	Pirate Moon Cluster	Tau	OC	04h 46m	+19° 07'	6.2	40'
118 Tau		Tau	MS	05h 29m	+25° 09'	5.5	5"
M 33	Triangulum Galaxy	Tri	Gx	01h 34m	+30° 40'	6.4	62'
Alp UMi	Polaris	UMi	MS	02h 51m	+89° 20'	2.0	18"

Chart 4

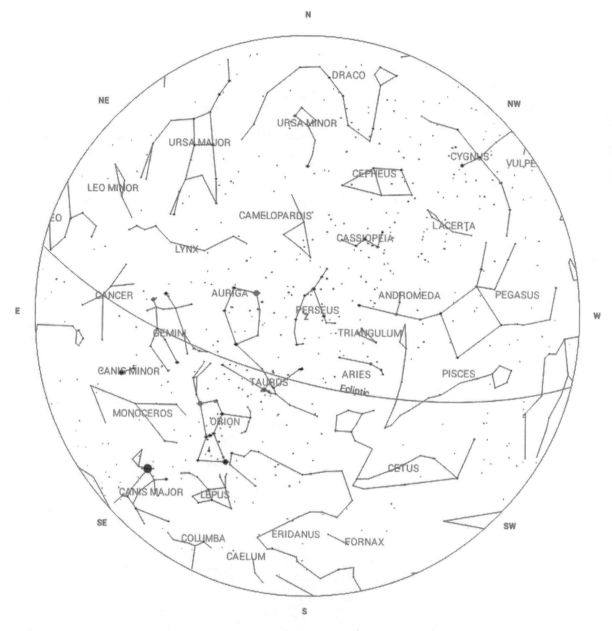

Designation	Name	Con.	Type	R.A.	Dec.	Mag	Size/Sep
Gam And	Almach	And	MS	02h 04m	+42° 20'	2.1	10"
Lam Ari		Ari	MS	01h 59m	+23° 41'	4.8	37"
Gam Ari	Mesarthim	Ari	MS	01h 54m	+19° 22'	4.6	8"
UU Aur		Aur	Var/CS	06h 36m	+38° 27'	5.3-6.5	N/A
The Aur		Aur	MS	06h 00m	+37° 13'	2.7	4"
M 37		Aur	OC	05h 52m	+32° 33'	6.2	14'
14 Aur		Aur	MS	05h 15m	+32° 41'	5.0	15"
Kemble 1	Kemble's Cascade	Cam	Ast	03h 57m	+63° 04'	5.0	180'

Designation	Name	Con.	Type	R.A.	Dec.	Mag	Size/Sep
NGC 1502	Jolly Roger Cluster	Cam	OC	04h 08m	+62° 20'	4.1	8'
NGC 457	Owl Cluster	Cas	OC	01h 20m	+58° 17'	5.1	20'
Iot Cas		Cas	MS	02h 29m	+67° 24'	4.5	7"
NGC 663		Cas	OC	01h 46m	+61° 14'	6.4	14'
Gam Cet	Kaffajidhma	Cet	MS	02h 43m	+03° 14'	3.5	3"
Omi Cet	Mira	Cet	Var	02h 19m	-02° 59'	2.0-10.1	N/A
32 Eri		Eri	MS	03h 54m	-02° 57'	4.4	7"
40 Eri	Keid	Eri	MS	04h 15m	-07° 39'	4.4	83"
M 35		Gem	OC	06h 09m	+24° 21'	5.6	25'
38 Gem		Gem	MS	06h 55m	+13° 11'	4.7	7"
Gam Lep		Lep	MS	05h 45m	-22° 27'	3.6	98"
R Lep	Hind's Crimson Star	Lep	Var/CS	05h 00m	-14° 48'	5.5-11.7	N/A
12 Lyn		Lyn	MS	06h 46m	+59° 27'	4.9	9"
Sig Ori		Ori	MS	05h 40m	-02° 36'	3.8	42"
Bet Ori	Rigel	Ori	MS	05h 14m	-08° 12'	0.2	10"
Alp Ori	Betelgeuse	Ori	Var	05h 55m	+07° 24'	0.4-1.3	N/A
Eta Ori		Ori	MS	05h 25m	-02° 24'	3.3	2"
Zet Ori	Alnitak	Ori	MS	05h 41m	-01° 57'	1.8	3"
23 Ori		Ori	MS	05h 23m	+03° 33'	5.0	32"
Iot Ori	Nair al Saif	Ori	MS	05h 35m	-05° 55'	2.8	11"
Collinder 70	Epsilon Orionis Cluster	Ori	OC	05h 36m	-01° 00'	0.4	150'
Struve 747		Ori	MS	05h 35m	-05° 55'	4.8	36"
Collinder 72		Ori	OC	05h 35m	-05° 55'	2.5	20'
Del Ori	Mintaka	Ori	MS	05h 33m	-00° 18'	2.1	53"
Lam Ori	Meissa	Ori	MC	05h 36m	+09° 56'	3.4	4"
M 42	Orion Nebula	Ori	Neb	05h 35m	-05° 23'	4.0	40'
NGC 1981	Coal Car Cluster	Ori	OC	05h 35m	-04° 26'	4.2	28'
Collinder 69	Lambda Orionis Cluster	Ori	OC	05h 35m	+09° 56'	2.8	70'
NGC 1528	m & m Double Cluster	Per	OC	04h 15m	+51° 13'	6.4	16'
Bet Per	Algol	Per	Var	03h 08m	+40° 57'	2.1-3.4	N/A
Eps Per		Per	MS	03h 58m	+40° 01'	2.9	9"
M 34		Per	OC	02h 42m	+42° 46'	5.8	35'
NGC 869/884	Double Cluster	Per	OC	02h 21m	+57° 08'	4.4	18'
Eta Per		Per	MS	02h 51m	+55° 54'	3.8	29"
Melotte 20	Alpha Persei Moving Cluster	Per	OC	03h 24m	+49° 52'	2.3	300'
Alp Psc	Alrisha	Psc	MS	02h 02m	+02° 46'	3.8	2"
Psi 1 Psc		Psc	MS	01h 06m	+21° 28'	5.3	30"
Zet Psc		Psc	MS	01h 14m	+07° 35'	5.2	23"
M 45	Pleiades	Tau	OC	03h 47m	+24° 07'	1.5	120'
Melotte 25	Hyades	Tau	OC	04h 27m	+15° 52'	0.8	330'
NGC 1647	Pirate Moon Cluster	Tau	OC	04h 46m	+19° 07'	6.2	40'
118 Tau		Tau	MS	05h 29m	+25° 09'	5.5	5"
M 33	Triangulum Galaxy	Tri	Gx	01h 34m	+30° 40'	6.4	62'
Alp UMi	Polaris	UMi	MS	02h 51m	+89° 20'	2.0	18"

Chart 5

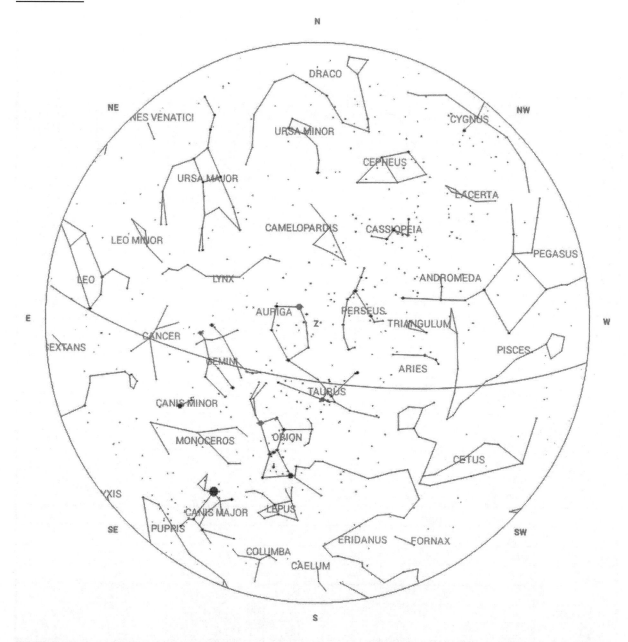

Designation	Name	Con.	Type	R.A.	Dec.	Mag	Size/Sep
Gam And	Almach	And	MS	02h 04m	+42° 20'	2.1	10"
The Aur		Aur	MS	06h 00m	+37° 13'	2.7	4"
14 Aur		Aur	MS	05h 15m	+32° 41'	5.0	15"
U Cam		Cam	Var/CS	03h 42m	+62° 39'	7.0-7.5	N/A
ST Cam		Cam	Var/CS	04h 51m	+68° 10'	7.0-8.4	N/A
Kemble 1	Kemble's Cascade	Cam	Ast	03h 57m	+63° 04'	5.0	180'
NGC 1502	Jolly Roger Cluster	Cam	OC	04h 08m	+62° 20'	4.1	8'
Iot Cas		Cas	MS	02h 29m	+67° 24'	4.5	7"

Designation	Name	Con.	Type	R.A.	Dec.	Mag	Size/Sep
32 Eri		Eri	MS	03h 54m	-02° 57'	4.4	7"
40 Eri	Keid	Eri	MS	04h 15m	-07° 39'	4.4	83"
M 35		Gem	OC	06h 09m	+24° 21'	5.6	25'
38 Gem		Gem	MS	06h 55m	+13° 11'	4.7	7"
Alp Gem	Castor	Gem	MS	07h 35m	+31° 53'	1.6	3"
Del Gem	Wasat	Gem	MS	07h 20m	+21° 59'	3.5	6"
Kap Gem		Gem	MS	07h 44m	+24° 24'	3.6	7"
Gam Lep		Lep	MS	05h 45m	-22° 27'	3.6	98"
R Lep	Hind's Crimson Star	Lep	Var/CS	05h 00m	-14° 48'	5.5-11.7	N/A
12 Lyn		Lyn	MS	06h 46m	+59° 27'	4.9	9"
19 Lyn		Lyn	MS	07h 23m	+55° 17'	5.8	215"
NGC 2353		Mon	OC	07h 15m	-10° 16'	5.2	18'
Bet Mon		Mon	MS	06h 29m	-07° 02'	3.8	7"
NGC 2237	Rosette Nebula	Mon	Neb	06h 32m	+04° 59'	5.5	70'
NGC 2244		Mon	OC	06h 32m	+04° 57'	5.2	29'
NGC 2264	Christmas Tree Cluster	Mon	OC	06h 41m	+09° 54'	4.1	39'
Eps Mon		Mon	MS	06h 24m	+04° 36'	4.3	13"
Sig Ori		Ori	MS	05h 40m	-02° 36'	3.8	42"
Bet Ori	Rigel	Ori	MS	05h 14m	-08° 12'	0.2	10"
Alp Ori	Betelgeuse	Ori	Var	05h 55m	+07° 24'	0.4-1.3	N/A
Eta Ori		Ori	MS	05h 25m	-02° 24'	3.3	2"
Zet Ori	Alnitak	Ori	MS	05h 41m	-01° 57'	1.8	3"
23 Ori		Ori	MS	05h 23m	+03° 33'	5.0	32"
Iot Ori	Nair al Saif	Ori	MS	05h 35m	-05° 55'	2.8	11"
W Ori		Ori	Var/CS	05h 05m	+01° 11'	6.2-7.0	N/A
Collinder 70	Epsilon Orionis Cluster	Ori	OC	05h 36m	-01° 00'	0.4	150'
Struve 747		Ori	MS	05h 35m	-05° 55'	4.8	36"
Collinder 72		Ori	OC	05h 35m	-05° 55'	2.5	20'
Del Ori	Mintaka	Ori	MS	05h 33m	-00° 18'	2.1	53"
Lam Ori	Meissa	Ori	MC	05h 36m	+09° 56'	3.4	4"
M 42	Orion Nebula	Ori	Neb	05h 35m	-05° 23'	4.0	40'
NGC 1981	Coal Car Cluster	Ori	OC	05h 35m	-04° 26'	4.2	28'
Collinder 69	Lambda Orionis Cluster	Ori	OC	05h 35m	+09° 56'	2.8	70'
BL Ori		Ori	Var/CS	06h 26m	+14° 43'	6.3-7.0	N/A
Bet Per	Algol	Per	Var	03h 08m	+40° 57'	2.1-3.4	N/A
Eps Per		Per	MS	03h 58m	+40° 01'	2.9	9"
M 34		Per	OC	02h 42m	+42° 46'	5.8	35'
NGC 869/884	Double Cluster	Per	OC	02h 21m	+57° 08'	4.4	18'
Eta Per		Per	MS	02h 51m	+55° 54'	3.8	29"
Melotte 20	Alpha Persei Moving Cluster	Per	OC	03h 24m	+49° 52'	2.3	300'
M 45	Pleiades	Tau	OC	03h 47m	+24° 07'	1.5	120'
Melotte 25	Hyades	Tau	OC	04h 27m	+15° 52'	0.8	330'
118 Tau		Tau	MS	05h 29m	+25° 09'	5.5	5"
Alp UMi	Polaris	UMi	MS	02h 51m	+89° 20'	2.0	18"

Chart 6

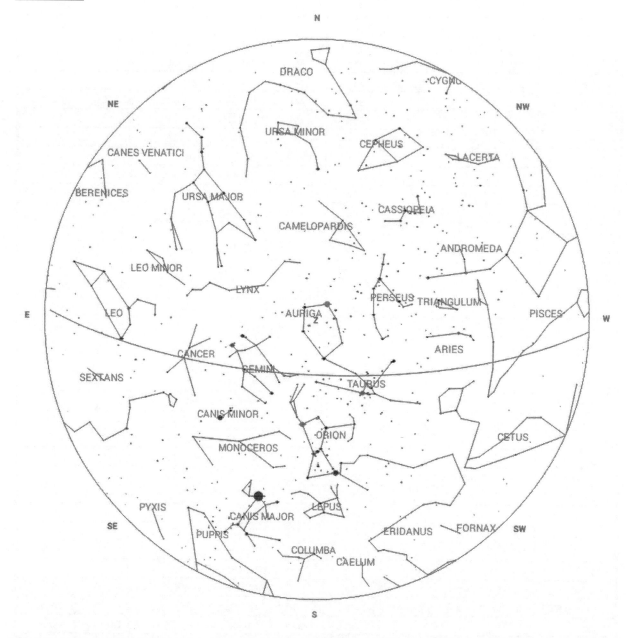

Designation	Name	Con.	Type	R.A.	Dec.	Mag	Size/Sep
UU Aur		Aur	Var/CS	06h 36m	+38° 27'	5.3-6.5	N/A
The Aur		Aur	MS	06h 00m	+37° 13'	2.7	4"
14 Aur		Aur	MS	05h 15m	+32° 41'	5.0	15"
Kemble 1	Kemble's Cascade	Cam	Ast	03h 57m	+63° 04'	5.0	180'
NGC 1502	Jolly Roger Cluster	Cam	OC	04h 08m	+62° 20'	4.1	8'
M 41		CMa	OC	05h 46m	-20° 45'	5.0	39'
Eps CMa	Adhara	CMa	MS	06h 59m	-28° 58'	1.5	8"
NGC 2362	Tau Canis Majoris Cluster	CMa	OC	07h 19m	-24° 57'	3.8	5'

Designation	Name	Con.	Type	R.A.	Dec.	Mag	Size/Sep
Iot Cnc		Cnc	MS	08h 47m	+28° 46'	4.0	30"
M 44	Praesepe	Cnc	OC	08h 40m	+19° 40'	3.9	70'
Zet Cnc	Tegmen	Cnc	MS	08h 12m	+17° 39'	4.7	6"
Phi2 Cnc		Cnc	MS	08h 27m	+26° 56'	5.6	5"
57 Cnc		Cnc	MS	08h 54m	+30° 35'	5.4	56"
X Cnc		Cnc	Var/CS	08h 55m	+17° 14'	5.6-7.5	N/A
32 Eri		Eri	MS	03h 54m	-02° 57'	4.4	7"
40 Eri	Keid	Eri	MS	04h 15m	-07° 39'	4.4	83"
M 35		Gem	OC	06h 09m	+24° 21'	5.6	25'
38 Gem		Gem	MS	06h 55m	+13° 11'	4.7	7"
Alp Gem	Castor	Gem	MS	07h 35m	+31° 53'	1.6	3"
Del Gem	Wasat	Gem	MS	07h 20m	+21° 59'	3.5	6"
Kap Gem		Gem	MS	07h 44m	+24° 24'	3.6	7"
Gam Lep		Lep	MS	05h 45m	-22° 27'	3.6	98"
R Lep	Hind's Crimson Star	Lep	Var/CS	05h 00m	-14° 48'	5.5-11.7	N/A
12 Lyn		Lyn	MS	06h 46m	+59° 27'	4.9	9"
19 Lyn		Lyn	MS	07h 23m	+55° 17'	5.8	215"
NGC 2353		Mon	OC	07h 15m	-10° 16'	5.2	18'
Bet Mon		Mon	MS	06h 29m	-07° 02'	3.8	7"
NGC 2237	Rosette Nebula	Mon	Neb	06h 32m	+04° 59'	5.5	70'
NGC 2244		Mon	OC	06h 32m	+04° 57'	5.2	29'
NGC 2264	Christmas Tree Cluster	Mon	OC	06h 41m	+09° 54'	4.1	39'
Eps Mon		Mon	MS	06h 24m	+04° 36'	4.3	13"
Sig Ori		Ori	MS	05h 40m	-02° 36'	3.8	42"
Bet Ori	Rigel	Ori	MS	05h 14m	-08° 12'	0.2	10"
Alp Ori	Betelgeuse	Ori	Var	05h 55m	+07° 24'	0.4-1.3	N/A
Eta Ori		Ori	MS	05h 25m	-02° 24'	3.3	2"
Zet Ori	Alnitak	Ori	MS	05h 41m	-01° 57'	1.8	3"
23 Ori		Ori	MS	05h 23m	+03° 33'	5.0	32"
Iot Ori	Nair al Saif	Ori	MS	05h 35m	-05° 55'	2.8	11"
Collinder 70	Epsilon Orionis Cluster	Ori	OC	05h 36m	-01° 00'	0.4	150'
Struve 747		Ori	MS	05h 35m	-05° 55'	4.8	36"
Collinder 72		Ori	OC	05h 35m	-05° 55'	2.5	20'
Del Ori	Mintaka	Ori	MS	05h 33m	-00° 18'	2.1	53"
Lam Ori	Meissa	Ori	MC	05h 36m	+09° 56'	3.4	4"
M 42	Orion Nebula	Ori	Neb	05h 35m	-05° 23'	4.0	40'
NGC 1981	Coal Car Cluster	Ori	OC	05h 35m	-04° 26'	4.2	28'
Collinder 69	Lambda Orionis Cluster	Ori	OC	05h 35m	+09° 56'	2.8	70'
Bet Per	Algol	Per	Var	03h 08m	+40° 57'	2.1-3.4	N/A
Eps Per		Per	MS	03h 58m	+40° 01'	2.9	9"
Melotte 20	Alpha Persei Moving Cluster	Per	OC	03h 24m	+49° 52'	2.3	300'
M 45	Pleiades	Tau	OC	03h 47m	+24° 07'	1.5	120'
Melotte 25	Hyades	Tau	OC	04h 27m	+15° 52'	0.8	330'
118 Tau		Tau	MS	05h 29m	+25° 09'	5.5	5"

Chart 7

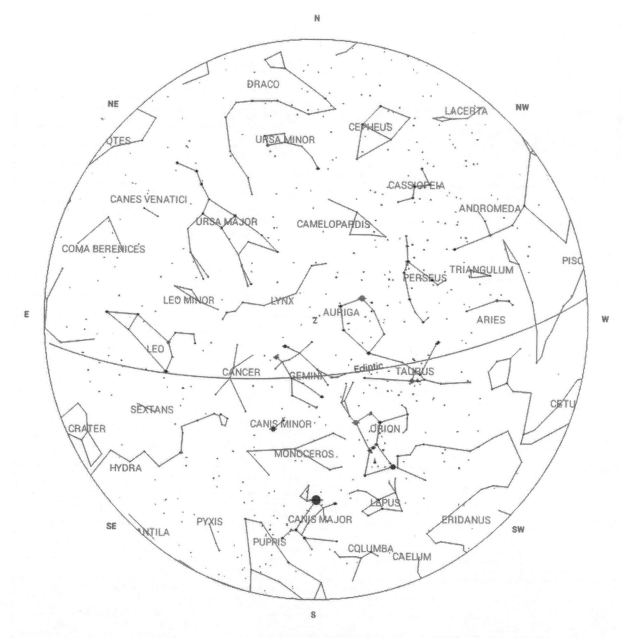

Designation	Name	Con.	Type	R.A.	Dec.	Mag	Size/Sep
14 Aur		Aur	MS	05h 15m	+32° 41'	5.0	15"
The Aur		Aur	MS	06h 00m	+37° 13'	2.7	4"
M 37		Aur	OC	05h 52m	+32° 33'	6.2	14'
UU Aur		Aur	Var/CS	06h 36m	+38° 27'	5.3-6.5	N/A
NGC 1502	Jolly Roger Cluster	Cam	OC	04h 08m	+62° 20'	4.1	8'
Eps CMa	Adhara	CMa	MS	06h 59m	-28° 58'	1.5	8"
M 41		CMa	OC	05h 46m	-20° 45'	5.0	39'
NGC 2362	Tau Canis Majoris Cluster	CMa	OC	07h 19m	-24° 57'	3.8	5'

Designation	Name	Con.	Type	R.A.	Dec.	Mag	Size/Sep
Zet Cnc	Tegmen	Cnc	MS	08h 12m	+17° 39'	4.7	6"
Phi2 Cnc		Cnc	MS	08h 27m	+26° 56'	5.6	5"
Iot Cnc		Cnc	MS	08h 47m	+28° 46'	4.0	30"
57 Cnc		Cnc	MS	08h 54m	+30° 35'	5.4	56"
M 44	Praesepe	Cnc	OC	08h 40m	+19° 40'	3.9	70'
X Cnc		Cnc	Var/CS	08h 55m	+17° 14'	5.6-7.5	N/A
38 Gem		Gem	MS	06h 55m	+13° 11'	4.7	7"
Del Gem	Wasat	Gem	MS	07h 20m	+21° 59'	3.5	6"
Alp Gem	Castor	Gem	MS	07h 35m	+31° 53'	1.6	3"
Kap Gem		Gem	MS	07h 44m	+24° 24'	3.6	7"
M 35		Gem	OC	06h 09m	+24° 21'	5.6	25'
R Leo	Peltier's Variable Star	Leo	Var	09h 48m	+11° 26'	4.4-10.5	N/A
Gam Lep		Lep	MS	05h 45m	-22° 27'	3.6	98"
R Lep	Hind's Crimson Star	Lep	Var/CS	05h 00m	-14° 48'	5.5-11.7	N/A
12 Lyn		Lyn	MS	06h 46m	+59° 27'	4.9	9"
19 Lyn		Lyn	MS	07h 23m	+55° 17'	5.8	215"
38 Lyn		Lyn	MS	09h 19m	+36° 48'	3.8	3"
Eps Mon		Mon	MS	06h 24m	+04° 36'	4.3	13"
Bet Mon		Mon	MS	06h 29m	-07° 02'	3.8	7"
NGC 2237	Rosette Nebula	Mon	Neb	06h 32m	+04° 59'	5.5	70'
NGC 2244		Mon	OC	06h 32m	+04° 57'	5.2	29'
NGC 2264	Christmas Tree Cluster	Mon	OC	06h 41m	+09° 54'	4.1	39'
NGC 2301	Hagrid's Dragon	Mon	OC	06h 52m	+00° 28'	6.3	14'
NGC 2353		Mon	OC	07h 15m	-10° 16'	5.2	18'
Lam Ori	Meissa	Ori	MC	05h 36m	+09° 56'	3.4	4"
Bet Ori	Rigel	Ori	MS	05h 14m	-08° 12'	0.2	10"
23 Ori		Ori	MS	05h 23m	+03° 33'	5.0	32"
Eta Ori		Ori	MS	05h 25m	-02° 24'	3.3	2"
Del Ori	Mintaka	Ori	MS	05h 33m	-00° 18'	2.1	53"
Iot Ori	Nair al Saif	Ori	MS	05h 35m	-05° 55'	2.8	11"
Struve 747		Ori	MS	05h 35m	-05° 55'	4.8	36"
Sig Ori		Ori	MS	05h 40m	-02° 36'	3.8	42"
Zet Ori	Alnitak	Ori	MS	05h 41m	-01° 57'	1.8	3"
M 42	Orion Nebula	Ori	Neb	05h 35m	-05° 23'	4.0	40'
Collinder 72		Ori	OC	05h 35m	-05° 55'	2.5	20'
NGC 1981	Coal Car Cluster	Ori	OC	05h 35m	-04° 26'	4.2	28'
Collinder 69	Lambda Orionis Cluster	Ori	OC	05h 35m	+09° 56'	2.8	70'
Collinder 70	Epsilon Orionis Cluster	Ori	OC	05h 36m	-01° 00'	0.4	150'
Alp Ori	Betelgeuse	Ori	Var	05h 55m	+07° 24'	0.4-1.3	N/A
M 47		Pup	OC	07h 37m	-14° 29'	4.3	25'
118 Tau		Tau	MS	05h 29m	+25° 09'	5.5	5"
Melotte 25	Hyades	Tau	OC	04h 27m	+15° 52'	0.8	330'
NGC 1647	Pirate Moon Cluster	Tau	OC	04h 46m	+19° 07'	6.2	40'

Chart 8

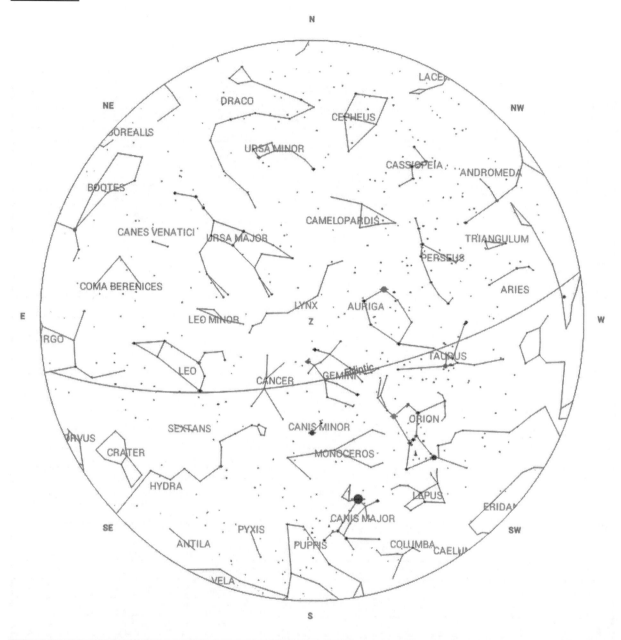

Designation	Name	Con.	Type	R.A.	Dec.	Mag	Size/Sep
14 Aur		Aur	MS	05h 15m	+32° 41'	5.0	15"
The Aur		Aur	MS	06h 00m	+37° 13'	2.7	4"
M 37		Aur	OC	05h 52m	+32° 33'	6.2	14'
UU Aur		Aur	Var/CS	06h 36m	+38° 27'	5.3-6.5	N/A
Eps CMa	Adhara	CMa	MS	06h 59m	-28° 58'	1.5	8"
M 41		CMa	OC	05h 46m	-20° 45'	5.0	39'
NGC 2362	Tau Canis Majoris Cluster	CMa	OC	07h 19m	-24° 57'	3.8	5'
Zet Cnc	Tegmen	Cnc	MS	08h 12m	+17° 39'	4.7	6"

Designation	Name	Con.	Type	R.A.	Dec.	Mag	Size/Sep
Phi 2 Cnc		Cnc	MS	08h 27m	+26° 56'	5.6	5"
Iot Cnc		Cnc	MS	08h 47m	+28° 46'	4.0	30"
57 Cnc		Cnc	MS	08h 54m	+30° 35'	5.4	56"
M 44	Praesepe	Cnc	OC	08h 40m	+19° 40'	3.9	70'
X Cnc		Cnc	Var/CS	08h 55m	+17° 14'	5.6-7.5	N/A
38 Gem		Gem	MS	06h 55m	+13° 11'	4.7	7"
Del Gem	Wasat	Gem	MS	07h 20m	+21° 59'	3.5	6"
Alp Gem	Castor	Gem	MS	07h 35m	+31° 53'	1.6	3"
Kap Gem		Gem	MS	07h 44m	+24° 24'	3.6	7"
M 35		Gem	OC	06h 09m	+24° 21'	5.6	25'
Eps Hya		Hya	MS	08h 47m	+06° 25'	3.4	3"
M 48		Hya	OC	08h 14m	-05° 45'	5.5	30'
U Hya		Hya	Var/CS	10h 38m	-13° 23'	4.8-6.5	N/A
R Leo	Peltier's Variable Star	Leo	Var	09h 48m	+11° 26'	4.4-10.5	N/A
Gam Leo	Algieba	Leo	MS	10h 20m	+19° 50'	2.0	5"
54 Leo		Leo	MS	10h 56m	+24° 45'	4.3	6"
12 Lyn		Lyn	MS	06h 46m	+59° 27'	4.9	9"
19 Lyn		Lyn	MS	07h 23m	+55° 17'	5.8	215"
38 Lyn		Lyn	MS	09h 19m	+36° 48'	3.8	3"
Eps Mon		Mon	MS	06h 24m	+04° 36'	4.3	13"
Bet Mon		Mon	MS	06h 29m	-07° 02'	3.8	7"
NGC 2237	Rosette Nebula	Mon	Neb	06h 32m	+04° 59'	5.5	70'
NGC 2244		Mon	OC	06h 32m	+04° 57'	5.2	29'
NGC 2264	Christmas Tree Cluster	Mon	OC	06h 41m	+09° 54'	4.1	39'
NGC 2301	Hagrid's Dragon	Mon	OC	06h 52m	+00° 28'	6.3	14'
NGC 2353		Mon	OC	07h 15m	-10° 16'	5.2	18'
Lam Ori	Meissa	Ori	MC	05h 36m	+09° 56'	3.4	4"
Bet Ori	Rigel	Ori	MS	05h 14m	-08° 12'	0.2	10"
23 Ori		Ori	MS	05h 23m	+03° 33'	5	32"
Eta Ori		Ori	MS	05h 25m	-02° 24'	3.3	2"
Del Ori	Mintaka	Ori	MS	05h 33m	-00° 18'	2.1	53"
Iot Ori	Nair al Saif	Ori	MS	05h 35m	-05° 55'	2.8	11"
Struve 747		Ori	MS	05h 35m	-05° 55'	4.8	36"
Sig Ori		Ori	MS	05h 40m	-02° 36'	3.8	42"
Zet Ori	Alnitak	Ori	MS	05h 41m	-01° 57'	1.8	3"
M 42	Orion Nebula	Ori	Neb	05h 35m	-05° 23'	4.0	40'
Collinder 72		Ori	OC	05h 35m	-05° 55'	2.5	20'
NGC 1981	Coal Car Cluster	Ori	OC	05h 35m	-04° 26'	4.2	28'
Collinder 69	Lambda Orionis Cluster	Ori	OC	05h 35m	+09° 56'	2.8	70'
Collinder 70	Epsilon Orionis Cluster	Ori	OC	05h 36m	-01° 00'	0.4	150'
Alp Ori	Betelgeuse	Ori	Var	05h 55m	+07° 24'	0.4-1.3	N/A
M 47		Pup	OC	07h 37m	-14° 29'	4.3	25'
118 Tau		Tau	MS	05h 29m	+25° 09'	5.5	5"

Chart 9

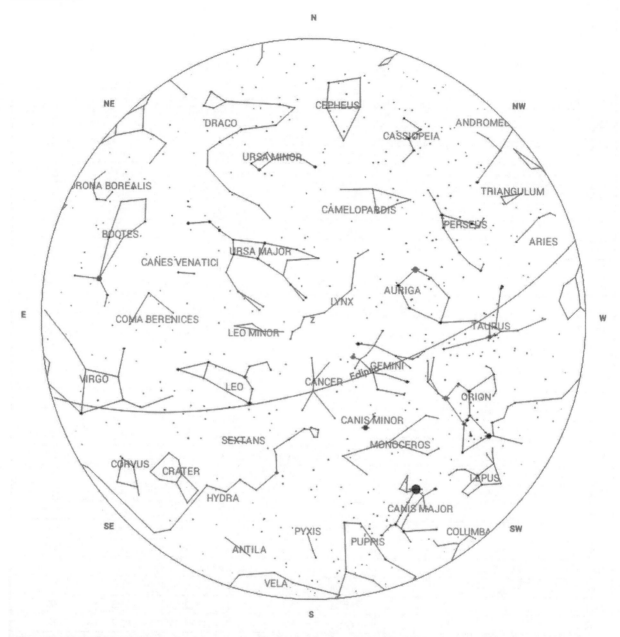

Designation	Name	Con.	Type	R.A.	Dec.	Mag	Size/Sep
The Aur		Aur	MS	06h 00m	+37° 13'	2.7	4"
UU Aur		Aur	Var/CS	06h 36m	+38° 27'	5.3-6.5	N/A
NGC 2403		Cam	Gx	07h 37m	+65° 36'	8.8	20'
Zet Cnc	Tegmen	Cnc	MS	08h 12m	+17° 39'	4.7	6"
Phi 2 Cnc		Cnc	MS	08h 27m	+26° 56'	5.6	5"
Iot Cnc		Cnc	MS	08h 47m	+28° 46'	4.0	30"
57 Cnc		Cnc	MS	08h 54m	+30° 35'	5.4	56"
M 44	Praesepe	Cnc	OC	08h 40m	+19° 40'	3.9	70'

Designation	Name	Con.	Type	R.A.	Dec.	Mag	Size/Sep
M 67		Cnc	OC	08h 51m	+11° 48'	7.4	25'
X Cnc		Cnc	Var/CS	08h 55m	+17° 14'	5.6-7.5	N/A
38 Gem		Gem	MS	06h 55m	+13° 11'	4.7	7"
Del Gem	Wasat	Gem	MS	07h 20m	+21° 59'	3.5	6"
Alp Gem	Castor	Gem	MS	07h 35m	+31° 53'	1.6	3"
Kap Gem		Gem	MS	07h 44m	+24° 24'	3.6	7"
M 35		Gem	OC	06h 09m	+24° 21'	5.6	25'
NGC 2392	Eskimo Nebula	Gem	PN	07h 29m	+20° 55'	8.6	47"
NGC 3242	Ghost of Jupiter	Hya	PN	10h 25m	-18° 39'	8.6	40"
Eps Hya		Hya	MS	08h 47m	+06° 25'	3.4	3"
M 48		Hya	OC	08h 14m	-05° 45'	5.5	30'
U Hya		Hya	Var/CS	10h 38m	-13° 23'	4.8-6.5	N/A
R Leo	Peltier's Variable Star	Leo	Var	09h 48m	+11° 26'	4.4-10.5	N/A
Gam Leo	Algieba	Leo	MS	10h 20m	+19° 50'	2.0	5"
54 Leo		Leo	MS	10h 56m	+24° 45'	4.3	6"
M 66		Leo	Gx	11h 20m	+13° 00'	9.7	9'
12 Lyn		Lyn	MS	06h 46m	+59° 27'	4.9	9"
19 Lyn		Lyn	MS	07h 23m	+55° 17'	5.8	215"
38 Lyn		Lyn	MS	09h 19m	+36° 48'	3.8	3"
Eps Mon		Mon	MS	06h 24m	+04° 36'	4.3	13"
NGC 3521		Leo	Gx	11h 06m	-00° 02'	9.9	10'
Bet Mon		Mon	MS	06h 29m	-07° 02'	3.8	7"
Iot Leo		Leo	MS	11h 24m	+10° 32'	3.9	2"
NGC 2237	Rosette Nebula	Mon	Neb	06h 32m	+04° 59'	5.5	70'
NGC 2244		Mon	OC	06h 32m	+04° 57'	5.2	29'
NGC 2264	Christmas Tree Cluster	Mon	OC	06h 41m	+09° 54'	4.1	39'
NGC 2301	Hagrid's Dragon	Mon	OC	06h 52m	+00° 28'	6.3	14'
M 50		Mon	OC	07h 03m	-08° 23'	7.2	14'
NGC 2353		Mon	OC	07h 15m	-10° 16'	5.2	18'
NGC 2506		Mon	OC	08h 00m	-10° 46'	8.9	12'
NGC 2175		Ori	Neb	06h 10m	+20° 29'	6.8	22'
NGC 2169	37 Cluster	Ori	OC	06h 09m	+13° 58'	7.0	5'
NGC 2467		Pup	Neb	07h 52m	-26° 26'	7.1	14'
M 47		Pup	OC	07h 37m	-14° 29'	4.3	25'
M 46		Pup	OC	07h 42m	-14° 48'	6.6	20'
M 93		Pup	OC	07h 45m	-23° 51'	6.5	10'
NGC 2539	Dish Cluster	Pup	OC	08h 11m	-12° 49'	8.0	9'
M 97	Owl Nebula	UMa	PN	11h 15m	+55° 01'	9.7	3'
Xi UMa	Alula Australis	UMa	MS	11h 18m	+31° 32'	4.4	2"
VY UMa		UMa	Var/CS	10h 45m	+67° 25'	5.9-6.5	N/A
M 81	Bode's Galaxy	UMa	Gx	09h 56m	+69° 04'	7.8	22'
M 82	Cigar Galaxy	UMa	Gx	09h 56m	+69° 41'	9.0	9'

Chart 10

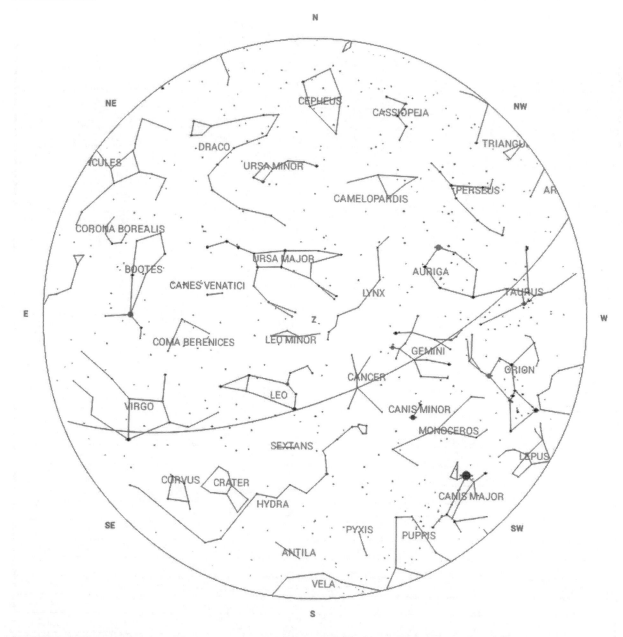

Designation	Name	Con.	Type	R.A.	Dec.	Mag	Size/Sep
NGC 2403		Cam	Gx	07h 37m	+65° 36'	8.8	20'
Zet Cnc	Tegmen	Cnc	MS	08h 12m	+17° 39'	4.7	6"
Phi 2 Cnc		Cnc	MS	08h 27m	+26° 56'	5.6	5"
Iot Cnc		Cnc	MS	08h 47m	+28° 46'	4.0	30"
57 Cnc		Cnc	MS	08h 54m	+30° 35'	5.4	56"
M 44	Praesepe	Cnc	OC	08h 40m	+19° 40'	3.9	70'
M 67		Cnc	OC	08h 51m	+11° 48'	7.4	25'
X Cnc		Cnc	Var/CS	08h 55m	+17° 14'	5.6-7.5	N/A
M 64	Black Eye Galaxy	Com	Gx	12h 57m	+21° 41'	9.3	10'

Designation	Name	Con.	Type	R.A.	Dec.	Mag	Size/Sep
Melotte 111	Coma Star Cluster	Com	OC	12h 25m	+26° 06'	2.9	120'
NGC 4725		Com	Gx	12h 50m	+25° 30'	9.9	10'
24 Com		Com	MS	12h 35m	+18° 23'	5.0	20"
M 106		CVn	Gx	12h 19m	+47° 18'	9.1	17'
Alp CVn	Cor Caroli	CVn	MS	12h 56m	+38° 19'	2.9	19"
M 94		CVn	Gx	12h 51m	+41° 07'	8.7	10'
2 CVn		CVn	MS	12h 16m	+40° 40'	5.7	11"
NGC 4656	Hook Galaxy	CVn	Gx	12h 44m	+32° 10'	9.7	9'
NGC 4449		CVn	Gx	12h 28m	+44° 06'	9.5	5'
Y CVn	La Superba	CVn	CS	12h 45m	+45° 26'	5.2-5.5	N/A
NGC 4490	Cocoon Galaxy	CVn	Gx	12h 31m	+41° 39'	9.8	6'
NGC 4631	Whale Galaxy	CVn	Gx	12h 42m	+32° 33'	9.5	13'
RY Dra		Dra	CS	12h 56m	+66° 00'	6.0-8.0	N/A
Del Gem	Wasat	Gem	MS	07h 20m	+21° 59'	3.5	6"
Alp Gem	Castor	Gem	MS	07h 35m	+31° 53'	1.6	3"
Kap Gem		Gem	MS	07h 44m	+24° 24'	3.6	7"
NGC 2392	Eskimo Nebula	Gem	PN	07h 29m	+20° 55'	8.6	47"
M 68		Hya	GC	12h 39m	-26° 45'	7.3	11'
NGC 3242	Ghost of Jupiter	Hya	PN	10h 25m	-18° 39'	8.6	40"
Eps Hya		Hya	MS	08h 47m	+06° 25'	3.4	3"
M 48		Hya	OC	08h 14m	-05° 45'	5.5	30'
U Hya		Hya	Var/CS	10h 38m	-13° 23'	4.8-6.5	N/A
R Leo	Peltier's Variable Star	Leo	Var	09h 48m	+11° 26'	4.4-10.5	N/A
Gam Leo	Algieba	Leo	MS	10h 20m	+19° 50'	2.0	5"
54 Leo		Leo	MS	10h 56m	+24° 45'	4.3	6"
M 66		Leo	Gx	11h 20m	+13° 00'	9.7	9'
NGC 3521		Leo	Gx	11h 06m	-00° 02'	9.9	10'
Iot Leo		Leo	MS	11h 24m	+10° 32'	3.9	2"
19 Lyn		Lyn	MS	07h 23m	+55° 17'	5.8	215"
38 Lyn		Lyn	MS	09h 19m	+36° 48'	3.8	3"
M 40	Winnecke 4	UMa	MS	12h 22m	+58° 05'	9.6	
M 97	Owl Nebula	UMa	PN	11h 15m	+55° 01'	9.7	3'
Xi UMa	Alula Australis	UMa	MS	11h 18m	+31° 32'	4.4	2"
VY UMa		UMa	Var/CS	10h 45m	+67° 25'	5.9-6.5	N/A
M 81	Bode's Galaxy	UMa	Gx	09h 56m	+69° 04'	7.8	22'
M 82	Cigar Galaxy	UMa	Gx	09h 56m	+69° 41'	9.0	9'
M 87		Vir	Gx	12h 31m	+12° 23'	9.6	8'
M 104	Sombrero Galaxy	Vir	Gx	12h 40m	-11° 37'	9.1	9'
M 49		Vir	Gx	12h 30m	+08° 00'	9.3	9'
M 60		Vir	Gx	12h 44m	+11° 33'	9.8	7'
M 86		Vir	Gx	12h 26m	+12° 57'	9.8	10'
Gam Vir	Porrima	Vir	MS	12h 42m	-01° 27'	2.7	2"
SS Vir		Vir	Var/CS	12h 25m	+00° 48'	6.0-9.6	6.0-9.6

Chart 11

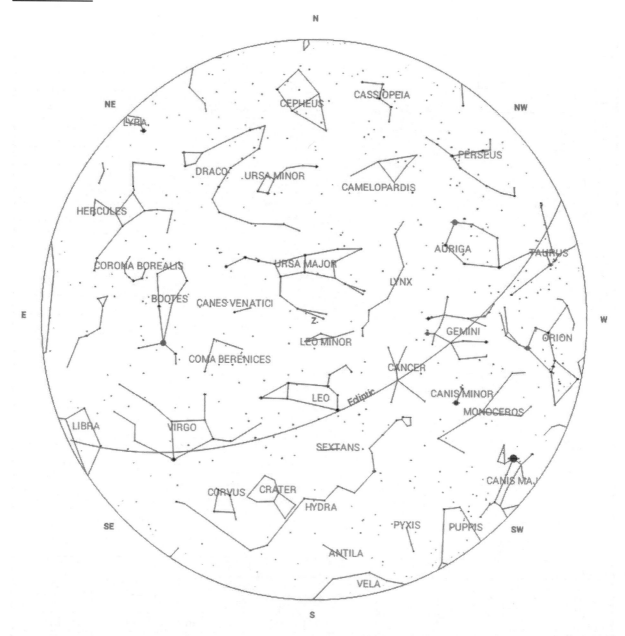

Designation	Name	Con.	Type	R.A.	Dec.	Mag	Size/Sep
Zet Cnc	Tegmen	Cnc	MS	08h 12m	+17° 39'	4.7	6"
Phi 2 Cnc		Cnc	MS	08h 27m	+26° 56'	5.6	5"
Iot Cnc		Cnc	MS	08h 47m	+28° 46'	4.0	30"
57 Cnc		Cnc	MS	08h 54m	+30° 35'	5.4	56"
M 44	Praesepe	Cnc	OC	08h 40m	+19° 40'	3.9	70'
M 67		Cnc	OC	08h 51m	+11° 48'	7.4	25'
X Cnc		Cnc	Var/CS	08h 55m	+17° 14'	5.6-7.5	N/A
M 53		Com	GC	13h 13m	+18° 10'	7.7	13'
M 64	Black Eye Galaxy	Com	Gx	12h 57m	+21° 41'	9.3	10'

Designation	Name	Con.	Type	R.A.	Dec.	Mag	Size/Sep
Melotte 111	Coma Star Cluster	Com	OC	12h 25m	+26° 06'	2.9	120'
NGC 4725		Com	Gx	12h 50m	+25° 30'	9.9	10'
24 Com		Com	MS	12h 35m	+18° 23'	5.0	20"
M 51	Whirlpool Galaxy	CVn	Gx	13h 30m	+47° 12'	8.7	10'
M 63	Sunflower Galaxy	CVn	Gx	13h 16m	+42° 02'	9.3	12'
M 106		CVn	Gx	12h 19m	+47° 18'	9.1	17'
M 3		CVn	GC	13h 42m	+28° 23'	6.3	18'
Alp CVn	Cor Caroli	CVn	MS	12h 56m	+38° 19'	2.9	19"
M 94		CVn	Gx	12h 51m	+41° 07'	8.7	10'
2 CVn		CVn	MS	12h 16m	+40° 40'	5.7	11"
NGC 4656	Hook Galaxy	CVn	Gx	12h 44m	+32° 10'	9.7	9'
NGC 4449		CVn	Gx	12h 28m	+44° 06'	9.5	5'
Y CVn	La Superba	CVn	CS	12h 45m	+45° 26'	5.2-5.5	N/A
NGC 4490	Cocoon Galaxy	CVn	Gx	12h 31m	+41° 39'	9.8	6'
NGC 4631	Whale Galaxy	CVn	Gx	12h 42m	+32° 33'	9.5	13'
RY Dra		Dra	CS	12h 56m	+66° 00'	6.0-8.0	N/A
M 83		Hya	Gx	13h 37m	-29° 52'	7.8	14'
M 68		Hya	GC	12h 39m	-26° 45'	7.3	11'
NGC 3242	Ghost of Jupiter	Hya	PN	10h 25m	-18° 39'	8.6	40"
Eps Hya		Hya	MS	08h 47m	+06° 25'	3.4	3"
M 48		Hya	OC	08h 14m	-05° 45'	5.5	30'
U Hya		Hya	Var/CS	10h 38m	-13° 23'	4.8-6.5	N/A
R Leo	Peltier's Variable Star	Leo	Var	09h 48m	+11° 26'	4.4-10.5	N/A
Gam Leo	Algieba	Leo	MS	10h 20m	+19° 50'	2.0	5"
54 Leo		Leo	MS	10h 56m	+24° 45'	4.3	6"
M 66		Leo	Gx	11h 20m	+13° 00'	9.7	9'
NGC 3521		Leo	Gx	11h 06m	-00° 02'	9.9	10'
Iot Leo		Leo	MS	11h 24m	+10° 32'	3.9	2"
38 Lyn		Lyn	MS	09h 19m	+36° 48'	3.8	3"
M 40	Winnecke 4	UMa	MS	12h 22m	+58° 05'	9.6	
M 97	Owl Nebula	UMa	PN	11h 15m	+55° 01'	9.7	3'
Zet UMa	Mizar & Alcor	UMa	MS	13h 24m	+54° 56'	2.1	711"
Xi UMa	Alula Australis	UMa	MS	11h 18m	+31° 32'	4.4	2"
VY UMa		UMa	Var/CS	10h 45m	+67° 25'	5.9-6.5	N/A
M 81	Bode's Galaxy	UMa	Gx	09h 56m	+69° 04'	7.8	22'
M 82	Cigar Galaxy	UMa	Gx	09h 56m	+69° 41'	9.0	9'
M 87		Vir	Gx	12h 31m	+12° 23'	9.6	8'
M 104	Sombrero Galaxy	Vir	Gx	12h 40m	-11° 37'	9.1	9'
M 49		Vir	Gx	12h 30m	+08° 00'	9.3	9'
M 60		Vir	Gx	12h 44m	+11° 33'	9.8	7'
The Vir		Vir	MS	13h 10m	-05° 32'	4.4	70"
M 86		Vir	Gx	12h 26m	+12° 57'	9.8	10'
Gam Vir	Porrima	Vir	MS	12h 42m	-01° 27'	2.7	2"
SS Vir		Vir	Var/CS	12h 25m	+00° 48'	6.0-9.6	6.0-9.6

Chart 12

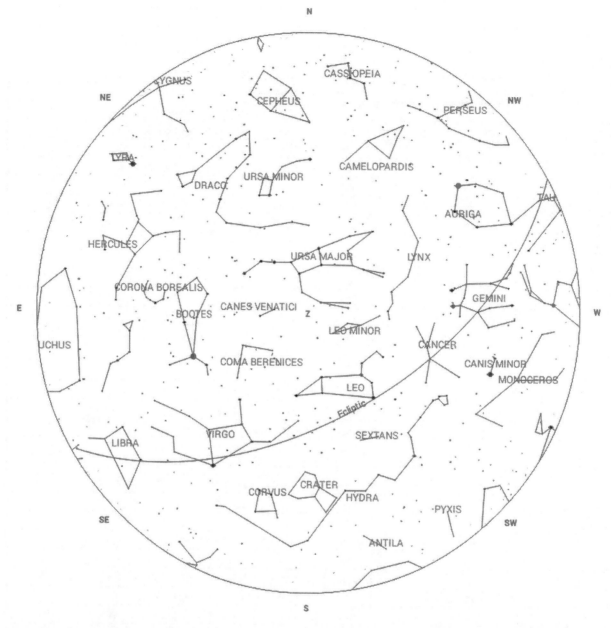

Designation	Name	Con.	Type	R.A.	Dec.	Mag	Size/Sep
Eps Boo	Izar	Boo	MS	14h 45m	+27° 04'	2.4	3"
Xi Boo		Boo	MS	14h 51m	+19° 06'	4.6	6"
Kap Boo	Asellus Tertius	Boo	MS	14h 14m	+51° 47'	4.5	13"
Pi Boo		Boo	MS	14h 41m	+16° 25'	4.5	6"
39 Boo		Boo	MS	14h 50m	+48° 43'	5.7	3"
Struve 1835		Boo	MS	14h 23m	+08° 27'	4.9	6"
M 53		Com	GC	13h 13m	+18° 10'	7.7	13'
M 64	Black Eye Galaxy	Com	Gx	12h 57m	+21° 41'	9.3	10'
Melotte 111	Coma Star Cluster	Com	OC	12h 25m	+26° 06'	2.9	120'

Designation	Name	Con.	Type	R.A.	Dec.	Mag	Size/Sep
24 Com		Com	MS	12h 35m	+18° 23'	5.0	20"
Del Crv	Algorab	Crv	MS	12h 30m	-16° 31'	5.7	24"
Struve 1669		Crv	MS	12h 41m	-13° 01'	5.2	5"
M 51	Whirlpool Galaxy	CVn	Gx	13h 30m	+47° 12'	8.7	10'
M 63	Sunflower Galaxy	CVn	Gx	13h 16m	+42° 02'	9.3	12'
M 106		CVn	Gx	12h 19m	+47° 18'	9.1	17'
M 3		CVn	GC	13h 42m	+28° 23'	6.3	18'
Alp CVn	Cor Caroli	CVn	MS	12h 56m	+38° 19'	2.9	19"
M 94		CVn	Gx	12h 51m	+41° 07'	8.7	10'
2 CVn		CVn	MS	12h 16m	+40° 40'	5.7	11"
NGC 4656	Hook Galaxy	CVn	Gx	12h 44m	+32° 10'	9.7	9'
NGC 4449		CVn	Gx	12h 28m	+44° 06'	9.5	5'
Y CVn	La Superba	CVn	CS	12h 45m	+45° 26'	5.2-5.5	N/A
NGC 4490	Cocoon Galaxy	CVn	Gx	12h 31m	+41° 39'	9.8	6'
NGC 4631	Whale Galaxy	CVn	Gx	12h 42m	+32° 33'	9.5	13'
RY Dra		Dra	CS	12h 56m	+66° 00'	6.0-8.0	N/A
M 83		Hya	Gx	13h 37m	-29° 52'	7.8	14'
54 Hya		Hya	MS	14h 46m	-25° 27'	5.2	9"
M 68		Hya	GC	12h 39m	-26° 45'	7.3	11'
NGC 3242	Ghost of Jupiter	Hya	PN	10h 25m	-18° 39'	8.6	40"
U Hya		Hya	Var/CS	10h 38m	-13° 23'	4.8-6.5	N/A
R Leo	Peltier's Variable Star	Leo	Var	09h 48m	+11° 26'	4.4-10.5	N/A
Gam Leo	Algieba	Leo	MS	10h 20m	+19° 50'	2.0	5"
54 Leo		Leo	MS	10h 56m	+24° 45'	4.3	6"
M 66		Leo	Gx	11h 20m	+13° 00'	9.7	9'
Iot Leo		Leo	MS	11h 24m	+10° 32'	3.9	2"
38 Lyn		Lyn	MS	09h 19m	+36° 48'	3.8	3"
M 101	Pinwheel Galaxy	UMa	Gx	14h 03m	+54° 21'	8.4	22'
M 40	Winnecke 4	UMa	MS	12h 22m	+58° 05'	9.6	
M 97	Owl Nebula	UMa	PN	11h 15m	+55° 01'	9.7	3'
Zet UMa	Mizar & Alcor	UMa	MS	13h 24m	+54° 56'	2.1	711"
Xi UMa	Alula Australis	UMa	MS	11h 18m	+31° 32'	4.4	2"
VY UMa		UMa	Var/CS	10h 45m	+67° 25'	5.9-6.5	N/A
M 81	Bode's Galaxy	UMa	Gx	09h 56m	+69° 04'	7.8	22'
M 82	Cigar Galaxy	UMa	Gx	09h 56m	+69° 41'	9.0	9'
M 87		Vir	Gx	12h 31m	+12° 23'	9.6	8'
M 104	Sombrero Galaxy	Vir	Gx	12h 40m	-11° 37'	9.1	9'
M 49		Vir	Gx	12h 30m	+08° 00'	9.3	9'
M 60		Vir	Gx	12h 44m	+11° 33'	9.8	7'
The Vir		Vir	MS	13h 10m	-05° 32'	4.4	70"
M 86		Vir	Gx	12h 26m	+12° 57'	9.8	10'
Gam Vir	Porrima	Vir	MS	12h 42m	-01° 27'	2.7	2"
SS Vir		Vir	Var/CS	12h 25m	+00° 48'	6.0-9.6	6.0-9.6

Chart 13

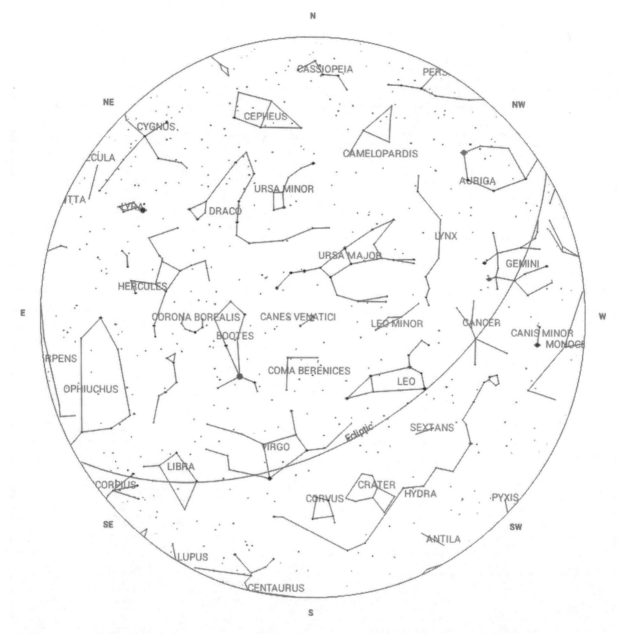

Designation	Name	Con.	Type	R.A.	Dec.	Mag	Size/Sep
Eps Boo	Izar	Boo	MS	14h 45m	+27° 04'	2.4	3"
Xi Boo		Boo	MS	14h 51m	+19° 06'	4.6	6"
Kap Boo	Asellus Tertius	Boo	MS	14h 14m	+51° 47'	4.5	13"
Pi Boo		Boo	MS	14h 41m	+16° 25'	4.5	6"
39 Boo		Boo	MS	14h 50m	+48° 43'	5.7	3"
Struve 1835		Boo	MS	14h 23m	+08° 27'	4.9	6"
Mu Boo	Alkalurops	Boo	MS	15h 24m	+37° 23'	4.3	108"
44 Boo		Boo	MS	15h 04m	+47° 39'	4.8	2"
M 53		Com	GC	13h 13m	+18° 10'	7.7	13'

Designation	Name	Con.	Type	R.A.	Dec.	Mag	Size/Sep
Del Boo		Boo	MS	15h 16m	+33° 19'	3.5	105"
Nu Boo		Boo	MS	15h 31m	+40° 50'	5.0	15'
M 64	Black Eye Galaxy	Com	Gx	12h 57m	+21° 41'	9.3	10'
Melotte 111	Coma Star Cluster	Com	OC	12h 25m	+26° 06'	2.9	120'
24 Com		Com	MS	12h 35m	+18° 23'	5.0	20"
Del Crv	Algorab	Crv	MS	12h 30m	-16° 31'	5.7	24"
Struve 1669		Crv	MS	12h 41m	-13° 01'	5.2	5"
M 51	Whirlpool Galaxy	CVn	Gx	13h 30m	+47° 12'	8.7	10'
M 63	Sunflower Galaxy	CVn	Gx	13h 16m	+42° 02'	9.3	12'
M 106		CVn	Gx	12h 19m	+47° 18'	9.1	17'
M 3		CVn	GC	13h 42m	+28° 23'	6.3	18'
Alp CVn	Cor Caroli	CVn	MS	12h 56m	+38° 19'	2.9	19"
M 94		CVn	Gx	12h 51m	+41° 07'	8.7	10'
2 CVn		CVn	MS	12h 16m	+40° 40'	5.7	11"
NGC 4656	Hook Galaxy	CVn	Gx	12h 44m	+32° 10'	9.7	9'
NGC 4449		CVn	Gx	12h 28m	+44° 06'	9.5	5'
Y CVn	La Superba	CVn	CS	12h 45m	+45° 26'	5.2-5.5	N/A
Zet CrB		CrB	MS	15h 39m	+36° 38'	4.6	6"
R CrB	Fade Out Star	CrB	Var	15h 49m	+28° 09'	5.7-14.8	N/A
NGC 4631	Whale Galaxy	CVn	Gx	12h 42m	+32° 33'	9.5	13'
RY Dra		Dra	CS	12h 56m	+66° 00'	6.0-8.0	N/A
M 83		Hya	Gx	13h 37m	-29° 52'	7.8	14'
54 Hya		Hya	MS	14h 46m	-25° 27'	5.2	9"
M 68		Hya	GC	12h 39m	-26° 45'	7.3	11'
NGC 3242	Ghost of Jupiter	Hya	PN	10h 25m	-18° 39'	8.6	40"
U Hya		Hya	Var/CS	10h 38m	-13° 23'	4.8-6.5	N/A
Gam Leo	Algieba	Leo	MS	10h 20m	+19° 50'	2.0	5"
54 Leo		Leo	MS	10h 56m	+24° 45'	4.3	6"
M 66		Leo	Gx	11h 20m	+13° 00'	9.7	9'
Iot Leo		Leo	MS	11h 24m	+10° 32'	3.9	2"
M 5		Ser	GC	15h 19m	+02° 05'	5.7	23'
Del Ser		Ser	MS	15h 35m	+10° 32'	4.2	4"
M 101	Pinwheel Galaxy	UMa	Gx	14h 03m	+54° 21'	8.4	22'
M 40	Winnecke 4	UMa	MS	12h 22m	+58° 05'	9.6	
M 97	Owl Nebula	UMa	PN	11h 15m	+55° 01'	9.7	3'
Zet UMa	Mizar & Alcor	UMa	MS	13h 24m	+54° 56'	2.1	711"
Xi UMa	Alula Australis	UMa	MS	11h 18m	+31° 32'	4.4	2"
VY UMa		UMa	Var/CS	10h 45m	+67° 25'	5.9-6.5	N/A
M 87		Vir	Gx	12h 31m	+12° 23'	9.6	8'
M 104	Sombrero Galaxy	Vir	Gx	12h 40m	-11° 37'	9.1	9'
M 49		Vir	Gx	12h 30m	+08° 00'	9.3	9'
The Vir		Vir	MS	13h 10m	-05° 32'	4.4	70"
Gam Vir	Porrima	Vir	MS	12h 42m	-01° 27'	2.7	2"
SS Vir		Vir	Var/CS	12h 25m	+00° 48'	6.0-9.6	6.0-9.6

Chart 14

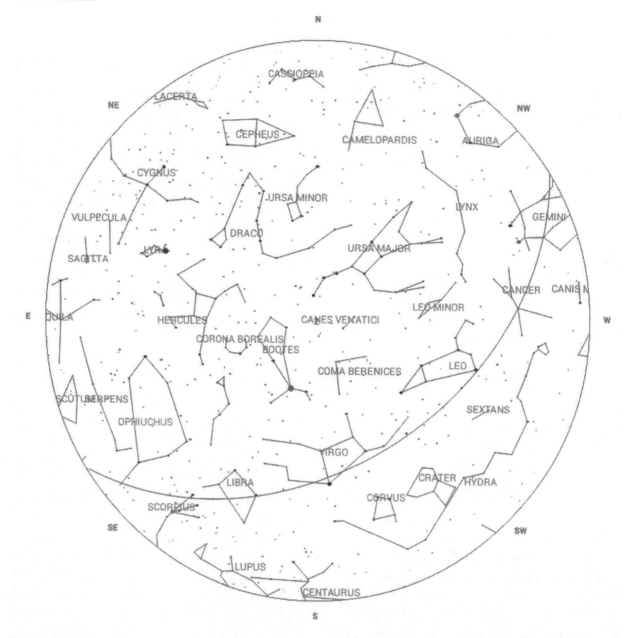

Designation	Name	Con.	Type	R.A.	Dec.	Mag	Size/Sep
Eps Boo	Izar	Boo	MS	14h 45m	+27° 04'	2.4	3"
Xi Boo		Boo	MS	14h 51m	+19° 06'	4.6	6"
Kap Boo	Asellus Tertius	Boo	MS	14h 14m	+51° 47'	4.5	13"
Pi Boo		Boo	MS	14h 41m	+16° 25'	4.5	6"
39 Boo		Boo	MS	14h 50m	+48° 43'	5.7	3"
Struve 1835		Boo	MS	14h 23m	+08° 27'	4.9	6"
Mu Boo	Alkalurops	Boo	MS	15h 24m	+37° 23'	4.3	108"
44 Boo		Boo	MS	15h 04m	+47° 39'	4.8	2"
Del Boo		Boo	MS	15h 16m	+33° 19'	3.5	105"

Designation	Name	Con.	Type	R.A.	Dec.	Mag	Size/Sep
Nu Boo		Boo	MS	15h 31m	+40° 50'	5.0	15'
M 53		Com	GC	13h 13m	+18° 10'	7.7	13'
M 64	Black Eye Galaxy	Com	Gx	12h 57m	+21° 41'	9.3	10'
Melotte 111	Coma Star Cluster	Com	OC	12h 25m	+26° 06'	2.9	120'
24 Com		Com	MS	12h 35m	+18° 23'	5.0	20"
Zet CrB		CrB	MS	15h 39m	+36° 38'	4.6	6"
Sig CrB		CrB	MS	16h 15m	+33° 52'	5.7	7"
R CrB	Fade Out Star	CrB	Var	15h 49m	+28° 09'	5.7-14.8	N/A
T CrB	Blaze Star	CrB	RN	16h 00m	+25° 55'	2.0-10.8	N/A
Del Crv	Algorab	Crv	MS	12h 30m	-16° 31'	5.7	24"
Struve 1669		Crv	MS	12h 41m	-13° 01'	5.2	5"
M 51	Whirlpool Galaxy	CVn	Gx	13h 30m	+47° 12'	8.7	10'
M 63	Sunflower Galaxy	CVn	Gx	13h 16m	+42° 02'	9.3	12'
M 106		CVn	Gx	12h 19m	+47° 18'	9.1	17'
M 3		CVn	GC	13h 42m	+28° 23'	6.3	18'
Alp CVn	Cor Caroli	CVn	MS	12h 56m	+38° 19'	2.9	19"
M 94		CVn	Gx	12h 51m	+41° 07'	8.7	10'
2 CVn		CVn	MS	12h 16m	+40° 40'	5.7	11"
NGC 4449		CVn	Gx	12h 28m	+44° 06'	9.5	5'
Y CVn	La Superba	CVn	CS	12h 45m	+45° 26'	5.2-5.5	N/A
NGC 4631	Whale Galaxy	CVn	Gx	12h 42m	+32° 33'	9.5	13'
RY Dra		Dra	CS	12h 56m	+66° 00'	6.0-8.0	N/A
16/17 Dra		Dra	MS	16h 36m	+52° 55'	5.1	90"
M 13	Keystone Cluster	Her	GC	16h 42m	+36° 27'	5.8	20'
Kap Her	Marfik	Her	MS	16h 41m	+31° 36'	5.0	28"
NGC 6229		Her	GC	16h 47m	+47° 32'	9.4	4'
Iot Leo		Leo	MS	11h 24m	+10° 32'	3.9	2"
Alp Lib	Zuben Elgenubi	Lib	MS	14h 51m	-16° 02'	2.8	230"
NGC 5897	Ghost Globular	Lib	GC	15h 17m	-21° 01'	8.4	11'
Bet Lib	The Emerald Star	Lib	*	15h 14m	-09° 23'	2.6	N/A
Struve 1962		Lib	MS	15h 39m	-08° 47'	5.4	12"
M 5		Ser	GC	15h 19m	+02° 05'	5.7	23'
Del Ser		Ser	MS	15h 35m	+10° 32'	4.2	4"
M 101	Pinwheel Galaxy	UMa	Gx	14h 03m	+54° 21'	8.4	22'
M 40	Winnecke 4	UMa	MS	12h 22m	+58° 05'	9.6	
Zet UMa	Mizar & Alcor	UMa	MS	13h 24m	+54° 56'	2.1	711"
Xi UMa	Alula Australis	UMa	MS	11h 18m	+31° 32'	4.4	2"
M 87		Vir	Gx	12h 31m	+12° 23'	9.6	8'
M 104	Sombrero Galaxy	Vir	Gx	12h 40m	-11° 37'	9.1	9'
M 49		Vir	Gx	12h 30m	+08° 00'	9.3	9'
The Vir		Vir	MS	13h 10m	-05° 32'	4.4	70"
Gam Vir	Porrima	Vir	MS	12h 42m	-01° 27'	2.7	2"
SS Vir		Vir	Var/CS	12h 25m	+00° 48'	6.0-9.6	6.0-9.6

Chart 15

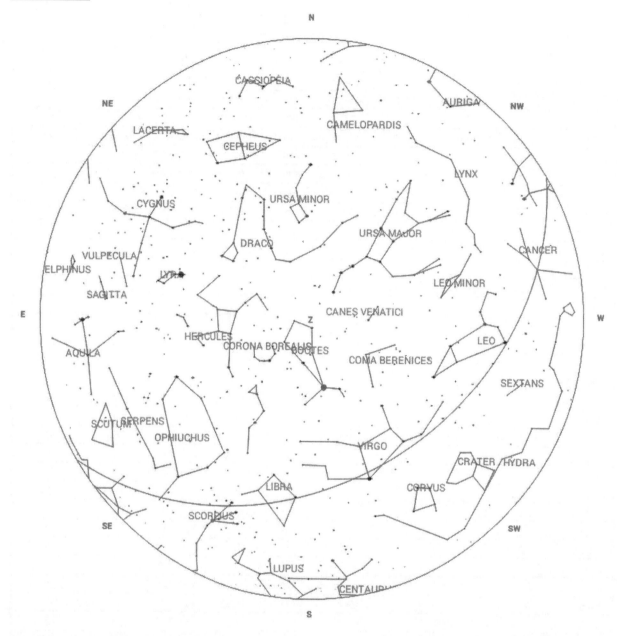

Designation	Name	Con.	Type	R.A.	Dec.	Mag	Size/Sep
Eps Boo	Izar	Boo	MS	14h 45m	+27° 04'	2.4	3"
Xi Boo		Boo	MS	14h 51m	+19° 06'	4.6	6"
Kap Boo	Asellus Tertius	Boo	MS	14h 14m	+51° 47'	4.5	13"
Pi Boo		Boo	MS	14h 41m	+16° 25'	4.5	6"
39 Boo		Boo	MS	14h 50m	+48° 43'	5.7	3"
Struve 1835		Boo	MS	14h 23m	+08° 27'	4.9	6"
Mu Boo	Alkalurops	Boo	MS	15h 24m	+37° 23'	4.3	108"
44 Boo		Boo	MS	15h 04m	+47° 39'	4.8	2"
Del Boo		Boo	MS	15h 16m	+33° 19'	3.5	105"

Designation	Name	Con.	Type	R.A.	Dec.	Mag	Size/Sep
Nu Boo		Boo	MS	15h 31m	+40° 50'	5.0	15'
M 53		Com	GC	13h 13m	+18° 10'	7.7	13'
Melotte 111	Coma Star Cluster	Com	OC	12h 25m	+26° 06'	2.9	120'
24 Com		Com	MS	12h 35m	+18° 23'	5.0	20"
Zet CrB		CrB	MS	15h 39m	+36° 38'	4.6	6"
Sig CrB		CrB	MS	16h 15m	+33° 52'	5.7	7"
R CrB	Fade Out Star	CrB	Var	15h 49m	+28° 09'	5.7-14.8	N/A
T CrB	Blaze Star	CrB	RN	16h 00m	+25° 55'	2.0-10.8	N/A
M 3		CVn	GC	13h 42m	+28° 23'	6.3	18'
Alp CVn	Cor Caroli	CVn	MS	12h 56m	+38° 19'	2.9	19"
2 CVn		CVn	MS	12h 16m	+40° 40'	5.7	11"
Y CVn	La Superba	CVn	CS	12h 45m	+45° 26'	5.2-5.5	N/A
RY Dra		Dra	CS	12h 56m	+66° 00'	6.0-8.0	N/A
16/17 Dra		Dra	MS	16h 36m	+52° 55'	5.1	90"
Nu Dra	Kuma	Dra	MS	17h 32m	+55° 10'	4.9	63"
Mu Dra		Dra	MS	17h 05m	+54° 28'	5.8	2"
Psi Dra		Dra	MS	17h 42m	+72° 09'	4.6	30"
M 13	Keystone Cluster	Her	GC	16h 42m	+36° 27'	5.8	20'
Kap Her	Marfik	Her	MS	16h 41m	+31° 36'	5.0	28"
M 92		Her	GC	17h 17m	+43° 08'	6.5	14'
Alp Her	Rasalgethi	Her	MS	17h 15m	+14° 23'	3.1	5"
Del Her	Sarin	Her	MS	17h 15m	+24° 50'	3.1	14"
Rho Her		Her	MS	17h 24m	+37° 09'	4.2	4"
Alp Lib	Zuben Elgenubi	Lib	MS	14h 51m	-16° 02'	2.8	230"
Bet Lib	The Emerald Star	Lib	*	15h 14m	-09° 23'	2.6	N/A
Struve 1962		Lib	MS	15h 39m	-08° 47'	5.4	12"
IC 4665	Summer Beehive	Oph	OC	17h 46m	+05° 43'	5.3	70'
M 10		Oph	GC	16h 57m	-04° 06'	6.6	20'
M 14		Oph	GC	17h 38m	-03° 15'	7.6	11'
M 12		Oph	GC	16h 47m	-01° 57'	6.1	16'
Rho Oph		Oph	MS	16h 26m	-23° 27'	4.6	3"
M 19		Oph	GC	17h 03m	-26° 16'	6.8	17'
M 62		Oph	GC	17h 01m	-30° 07'	6.4	15'
36 Oph		Oph	MS	17h 15m	-26° 36'	4.3	730"
Omi Oph		Oph	MS	17h 18m	-24° 17'	5.1	10"
61 Oph		Oph	MS	17h 45m	+02° 35'	6.2	21"
M 5		Ser	GC	15h 19m	+02° 05'	5.7	23'
Del Ser		Ser	MS	15h 35m	+10° 32'	4.2	4"
Zet UMa	Mizar & Alcor	UMa	MS	13h 24m	+54° 56'	2.1	711"
The Vir		Vir	MS	13h 10m	-05° 32'	4.4	70"
Gam Vir	Porrima	Vir	MS	12h 42m	-01° 27'	2.7	2"
SS Vir		Vir	Var/CS	12h 25m	+00° 48'	6.0-9.6	6.0-9.6

Chart 16

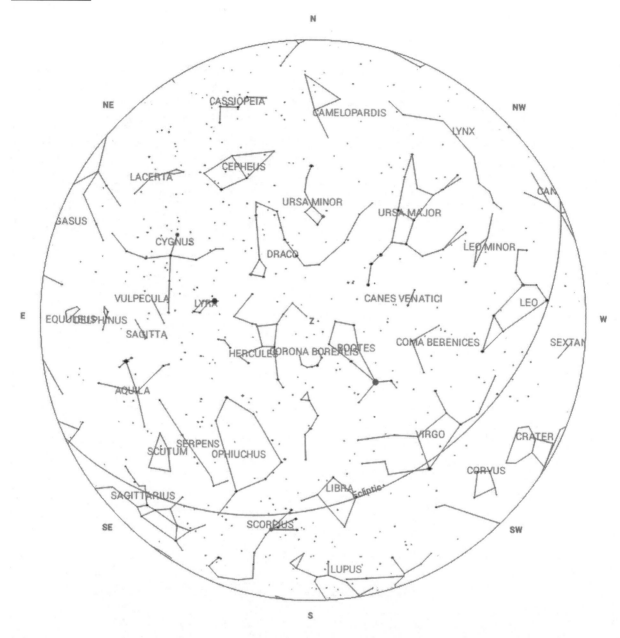

Designation	Name	Con.	Type	R.A.	Dec.	Mag	Size/Sep
Eps Boo	Izar	Boo	MS	14h 45m	+27° 04'	2.4	3"
Xi Boo		Boo	MS	14h 51m	+19° 06'	4.6	6"
Kap Boo	Asellus Tertius	Boo	MS	14h 14m	+51° 47'	4.5	13"
Pi Boo		Boo	MS	14h 41m	+16° 25'	4.5	6"
39 Boo		Boo	MS	14h 50m	+48° 43'	5.7	3"
Struve 1835		Boo	MS	14h 23m	+08° 27'	4.9	6"
Mu Boo	Alkalurops	Boo	MS	15h 24m	+37° 23'	4.3	108"
44 Boo		Boo	MS	15h 04m	+47° 39'	4.8	2"
Del Boo		Boo	MS	15h 16m	+33° 19'	3.5	105"

Designation	Name	Con.	Type	R.A.	Dec.	Mag	Size/Sep
Nu Boo		Boo	MS	15h 31m	+40° 50'	5.0	15'
Zet CrB		CrB	MS	15h 39m	+36° 38'	4.6	6"
Sig CrB		CrB	MS	16h 15m	+33° 52'	5.7	7"
R CrB	Fade Out Star	CrB	Var	15h 49m	+28° 09'	5.7-14.8	N/A
T CrB	Blaze Star	CrB	RN	16h 00m	+25° 55'	2.0-10.8	N/A
M 3		CVn	GC	13h 42m	+28° 23'	6.3	18'
16/17 Dra		Dra	MS	16h 36m	+52° 55'	5.1	90"
Nu Dra	Kuma	Dra	MS	17h 32m	+55° 10'	4.9	63"
Mu Dra		Dra	MS	17h 05m	+54° 28'	5.8	2"
Psi Dra		Dra	MS	17h 42m	+72° 09'	4.6	30"
39 Dra		Dra	MS	18h 24m	+58° 48'	5.0	89"
40/41 Dra		Dra	MS	18h 00m	+80° 00'	5.7	222"
M 13	Keystone Cluster	Her	GC	16h 42m	+36° 27'	5.8	20'
Kap Her	Marfik	Her	MS	16h 41m	+31° 36'	5.0	28"
M 92		Her	GC	17h 17m	+43° 08'	6.5	14'
Alp Her	Rasalgethi	Her	MS	17h 15m	+14° 23'	3.1	5"
Del Her	Sarin	Her	MS	17h 15m	+24° 50'	3.1	14"
Rho Her		Her	MS	17h 24m	+37° 09'	4.2	4"
95 Her		Her	MS	18h 02m	+21° 36'	4.3	6"
100 Her		Her	MS	18h 08m	+26° 06'	5.8	14"
Alp Lib	Zuben Elgenubi	Lib	MS	14h 51m	-16° 02'	2.8	230"
Bet Lib	The Emerald Star	Lib	*	15h 14m	-09° 23'	2.6	N/A
Struve 1962		Lib	MS	15h 39m	-08° 47'	5.4	12"
Eps Lyr	The Double Double	Lyr	MS	18h 44m	+39° 40'	4.7	3"
Bet Lyr	Sheliak	Lyr	MS	18h 50m	+33° 22'	3.2	86"
Del Lyr		Lyr	MS	18h 54m	+36° 58'	4.2	630"
Zet Lyr		Lyr	MS	18h 45m	+37° 36'	4.4	44"
IC 4665	Summer Beehive	Oph	OC	17h 46m	+05° 43'	5.3	70'
M 10		Oph	GC	16h 57m	-04° 06'	6.6	20'
M 12		Oph	GC	16h 47m	-01° 57'	6.1	16'
Rho Oph		Oph	MS	16h 26m	-23° 27'	4.6	3"
M 62		Oph	GC	17h 01m	-30° 07'	6.4	15'
36 Oph		Oph	MS	17h 15m	-26° 36'	4.3	730"
Omi Oph		Oph	MS	17h 18m	-24° 17'	5.1	10"
61 Oph		Oph	MS	17h 45m	+02° 35'	6.2	21"
NGC 6633	Tweedledum Cluster	Oph	OC	18h 27m	+06° 31'	5.6	20'
70 Oph		Oph	MS	18h 06m	+02° 30'	4.2	4"
IC 4756		Ser	OC	18h 39m	+05° 27'	5.4	39'
M 16	Eagle Nebula	Ser	Neb	18h 19m	-13° 49'	6.0	9'
The Ser		Ser	MS	18h 56m	+04° 12'	4.3	22"
M 5		Ser	GC	15h 19m	+02° 05'	5.7	23'
Del Ser		Ser	MS	15h 35m	+10° 32'	4.2	4"
Zet UMa	Mizar & Alcor	UMa	MS	13h 24m	+54° 56'	2.1	711"
The Vir		Vir	MS	13h 10m	-05° 32'	4.4	70"

Chart 17

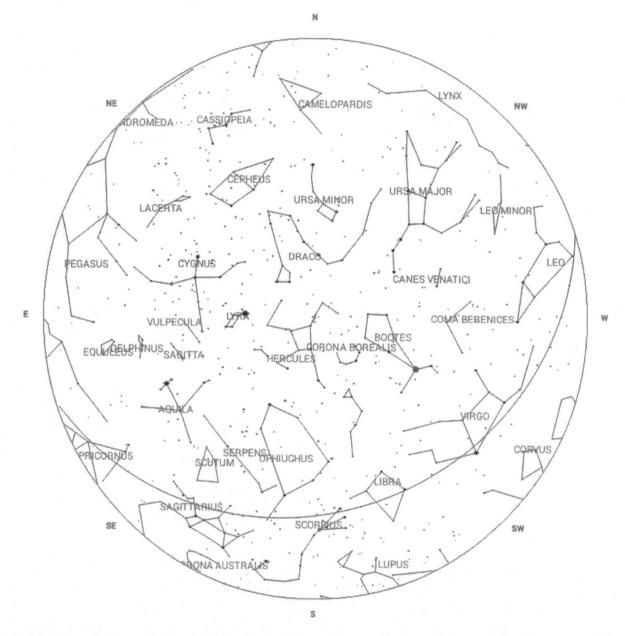

Designation	Name	Con.	Type	R.A.	Dec.	Mag	Size/Sep
15 Aql		Aql	MS	19h 06m	-04° 00'	5.4	39"
57 Aql		Aql	MS	19h 55m	-08° 14'	5.7	36"
Eps Boo	Izar	Boo	MS	14h 45m	+27° 04'	2.4	3"
Xi Boo		Boo	MS	14h 51m	+19° 06'	4.6	6"
Kap Boo	Asellus Tertius	Boo	MS	14h 14m	+51° 47'	4.5	13"
Pi Boo		Boo	MS	14h 41m	+16° 25'	4.5	6"
39 Boo		Boo	MS	14h 50m	+48° 43'	5.7	3"
Struve 1835		Boo	MS	14h 23m	+08° 27'	4.9	6"
Mu Boo	Alkalurops	Boo	MS	15h 24m	+37° 23'	4.3	108"

Designation	Name	Con.	Type	R.A.	Dec.	Mag	Size/Sep
44 Boo		Boo	MS	15h 04m	+47° 39'	4.8	2"
Del Boo		Boo	MS	15h 16m	+33° 19'	3.5	105"
Nu Boo		Boo	MS	15h 31m	+40° 50'	5.0	15'
Zet CrB		CrB	MS	15h 39m	+36° 38'	4.6	6"
Sig CrB		CrB	MS	16h 15m	+33° 52'	5.7	7"
Bet Cyg	Albireo	Cyg	MS	19h 31m	+27° 58'	3.1	34"
Del Cyg		Cyg	MS	19h 45m	+45° 08'	2.9	2"
16/17 Dra		Dra	MS	16h 36m	+52° 55'	5.1	90"
Nu Dra	Kuma	Dra	MS	17h 32m	+55° 10'	4.9	63"
Psi Dra		Dra	MS	17h 42m	+72° 09'	4.6	30"
39 Dra		Dra	MS	18h 24m	+58° 48'	5.0	89"
40/41 Dra		Dra	MS	18h 00m	+80° 00'	5.7	222"
Eps Dra		Dra	MS	19h 48m	+70° 16'	3.8	3"
Kap Her	Marfik	Her	MS	16h 41m	+31° 36'	5.0	28"
Alp Her	Rasalgethi	Her	MS	17h 15m	+14° 23'	3.1	5"
Del Her	Sarin	Her	MS	17h 15m	+24° 50'	3.1	14"
Rho Her		Her	MS	17h 24m	+37° 09'	4.2	4"
95 Her		Her	MS	18h 02m	+21° 36'	4.3	6"
Alp Lib	Zuben Elgenubi	Lib	MS	14h 51m	-16° 02'	2.8	230"
Bet Lib	The Emerald Star	Lib	*	15h 14m	-09° 23'	2.6	N/A
Struve 1962		Lib	MS	15h 39m	-08° 47'	5.4	12"
Eps Lyr	The Double Double	Lyr	MS	18h 44m	+39° 40'	4.7	3"
Bet Lyr	Sheliak	Lyr	MS	18h 50m	+33° 22'	3.2	86"
Del Lyr		Lyr	MS	18h 54m	+36° 58'	4.2	630"
Zet Lyr		Lyr	MS	18h 45m	+37° 36'	4.4	44"
IC 4665	Summer Beehive	Oph	OC	17h 46m	+05° 43'	5.3	70'
Rho Oph		Oph	MS	16h 26m	-23° 27'	4.6	3"
36 Oph		Oph	MS	17h 15m	-26° 36'	4.3	730"
Omi Oph		Oph	MS	17h 18m	-24° 17'	5.1	10"
NGC 6633	Tweedledum Cluster	Oph	OC	18h 27m	+06° 31'	5.6	20'
70 Oph		Oph	MS	18h 06m	+02° 30'	4.2	4"
M 6	Butterfly Cluster	Sco	OC	17h 40m	-32° 15'	4.6	20'
M 7		Sco	OC	17h 54m	-34° 48'	3.3	80'
M 4		Sco	GC	16h 24m	-26° 32'	5.4	36'
Alp Sco	Antares	Sco	MS	16h 29m	-26° 26'	1.0	3"
Bet Sco	Graffias	Sco	MS	16h 05m	-19° 48'	2.6	14"
Nu Sco	Jabbah	Sco	MS	16h 12m	-19° 28'	4.0	2"
Xi Sco		Sco	MS	16h 04m	-11° 22'	4.2	8"
Struve 1999		Sco	MS	16h 04m	-11° 22'	4.2	12"
IC 4756		Ser	OC	18h 39m	+05° 27'	5.4	39'
The Ser		Ser	MS	18h 56m	+04° 12'	4.3	22"
M 5		Ser	GC	15h 19m	+02° 05'	5.7	23'
Del Ser		Ser	MS	15h 35m	+10° 32'	4.2	4"
Collinder 399	Coathanger	Vul	Ast	19h 25m	+20° 11'	4.8	89'

Chart 18

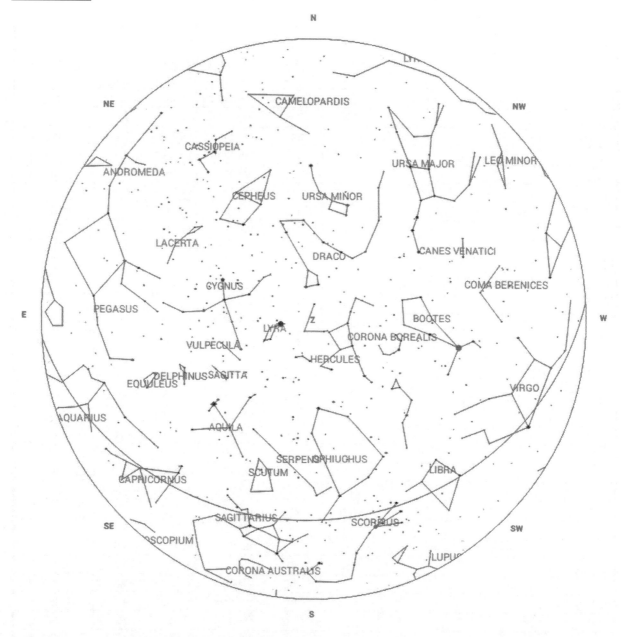

Designation	Name	Con.	Type	R.A.	Dec.	Mag	Size/Sep
15 Aql		Aql	MS	19h 06m	-04° 00'	5.4	39"
57 Aql		Aql	MS	19h 55m	-08° 14'	5.7	36"
Mu Boo	Alkalurops	Boo	MS	15h 24m	+37° 23'	4.3	108"
44 Boo		Boo	MS	15h 04m	+47° 39'	4.8	2"
Del Boo		Boo	MS	15h 16m	+33° 19'	3.5	105"
Nu Boo		Boo	MS	15h 31m	+40° 50'	5.0	15'
Zet CrB		CrB	MS	15h 39m	+36° 38'	4.6	6"
Sig CrB		CrB	MS	16h 15m	+33° 52'	5.7	7"
Bet Cyg	Albireo	Cyg	MS	19h 31m	+27° 58'	3.1	34"

Designation	Name	Con.	Type	R.A.	Dec.	Mag	Size/Sep
Del Cyg		Cyg	MS	19h 45m	+45° 08'	2.9	2"
NGC 7000	North American Nebula	Cyg	Neb	20h 59m	+44° 22'	4.0	120'
Gam Del		Del	MS	20h 47m	+16° 07'	3.9	9"
16/17 Dra		Dra	MS	16h 36m	+52° 55'	5.1	90"
Nu Dra	Kuma	Dra	MS	17h 32m	+55° 10'	4.9	63"
Psi Dra		Dra	MS	17h 42m	+72° 09'	4.6	30"
39 Dra		Dra	MS	18h 24m	+58° 48'	5.0	89"
40/41 Dra		Dra	MS	18h 00m	+80° 00'	5.7	222"
Eps Dra		Dra	MS	19h 48m	+70° 16'	3.8	3"
Eps Equ		Equ	MS	20h 59m	+04° 18'	5.2	11"
Kap Her	Marfik	Her	MS	16h 41m	+31° 36'	5.0	28"
Alp Her	Rasalgethi	Her	MS	17h 15m	+14° 23'	3.1	5"
Del Her	Sarin	Her	MS	17h 15m	+24° 50'	3.1	14"
Rho Her		Her	MS	17h 24m	+37° 09'	4.2	4"
95 Her		Her	MS	18h 02m	+21° 36'	4.3	6"
Eps Lyr	The Double Double	Lyr	MS	18h 44m	+39° 40'	4.7	3"
Bet Lyr	Sheliak	Lyr	MS	18h 50m	+33° 22'	3.2	86"
Del Lyr		Lyr	MS	18h 54m	+36° 58'	4.2	630"
Zet Lyr		Lyr	MS	18h 45m	+37° 36'	4.4	44"
IC 4665	Summer Beehive	Oph	OC	17h 46m	+05° 43'	5.3	70'
Rho Oph		Oph	MS	16h 26m	-23° 27'	4.6	3"
36 Oph		Oph	MS	17h 15m	-26° 36'	4.3	730"
Omi Oph		Oph	MS	17h 18m	-24° 17'	5.1	10"
NGC 6633	Tweedledum Cluster	Oph	OC	18h 27m	+06° 31'	5.6	20'
70 Oph		Oph	MS	18h 06m	+02° 30'	4.2	4"
M 6	Butterfly Cluster	Sco	OC	17h 40m	-32° 15'	4.6	20'
M 7		Sco	OC	17h 54m	-34° 48'	3.3	80'
M 4		Sco	GC	16h 24m	-26° 32'	5.4	36'
Alp Sco	Antares	Sco	MS	16h 29m	-26° 26'	1.0	3"
Bet Sco	Graffias	Sco	MS	16h 05m	-19° 48'	2.6	14"
Nu Sco	Jabbah	Sco	MS	16h 12m	-19° 28'	4.0	2"
Xi Sco		Sco	MS	16h 04m	-11° 22'	4.2	8"
Struve 1999		Sco	MS	16h 04m	-11° 22'	4.2	12"
IC 4756		Ser	OC	18h 39m	+05° 27'	5.4	39'
The Ser		Ser	MS	18h 56m	+04° 12'	4.3	22"
M 5		Ser	GC	15h 19m	+02° 05'	5.7	23'
Del Ser		Ser	MS	15h 35m	+10° 32'	4.2	4"
The Sge		Sge	MS	20h 10m	+20° 55'	4.6	84"
M 24	Sagittarius Star Cloud	Sgr	OC	18h 18m	-18° 24'	3.1	90'
M 22		Sgr	GC	18h 36m	-23° 54'	5.2	32'
M 8	Lagoon Nebula	Sgr	Neb	18h 04m	-24° 23'	5.0	17'
Collinder 399	Coathanger	Vul	Ast	19h 25m	+20° 11'	4.8	89'
NGC 6885		Vul	OC	20h 12m	+26° 29'	5.7	20'

Chart 19

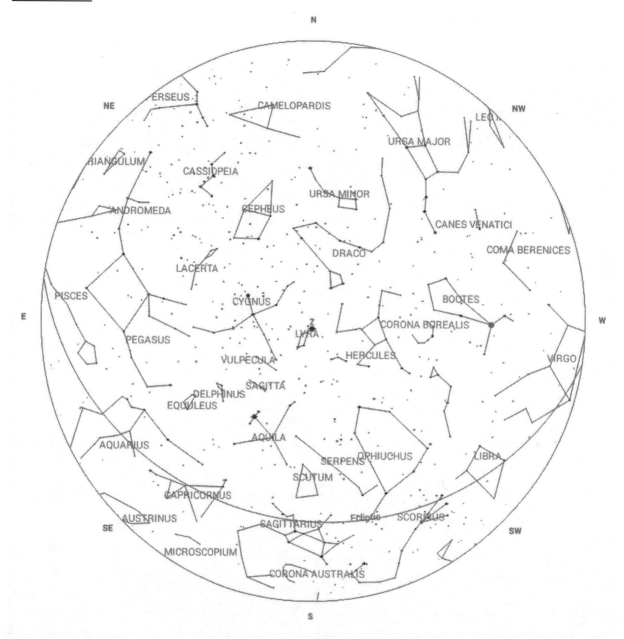

Designation	Name	Con.	Type	R.A.	Dec.	Mag	Size/Sep
15 Aql		Aql	MS	19h 06m	-04° 00'	5.4	39"
57 Aql		Aql	MS	19h 55m	-08° 14'	5.7	36"
Bet Cyg	Albireo	Cyg	MS	19h 31m	+27° 58'	3.1	34"
Del Cyg		Cyg	MS	19h 45m	+45° 08'	2.9	2"
16 Cyg		Cyg	MS	19h 42m	+50° 32'	5.9	39"
61 Cyg		Cyg	MS	21h 07m	+38° 44'	5.2	29"
M 39		Cyg	OC	21h 32m	+48° 26'	5.3	29'
Mu Cyg		Cyg	MS	21h 44m	+28° 45'	4.5	200"

Designation	Name	Con.	Type	R.A.	Dec.	Mag	Size/Sep
V460		Cyg	Var/CS	21h 42m	+35° 31'	5.6-7.0	N/A
NGC 7000	North American Nebula	Cyg	Neb	20h 59m	+44° 22'	4.0	120'
	Northern Coalsack	Cyg	DN	20h 41m	+43° 00'	6.0	60'
Gam Del		Del	MS	20h 47m	+16° 07'	3.9	9"
16/17 Dra		Dra	MS	16h 36m	+52° 55'	5.1	90"
Nu Dra	Kuma	Dra	MS	17h 32m	+55° 10'	4.9	63"
Mu Dra		Dra	MS	17h 05m	+54° 28'	5.8	2"
Psi Dra		Dra	MS	17h 42m	+72° 09'	4.6	30"
39 Dra		Dra	MS	18h 24m	+58° 48'	5.0	89"
40/41 Dra		Dra	MS	18h 00m	+80° 00'	5.7	222"
Eps Dra		Dra	MS	19h 48m	+70° 16'	3.8	3"
Eps Equ		Equ	MS	20h 59m	+04° 18'	5.2	11"
M 13	Keystone Cluster	Her	GC	16h 42m	+36° 27'	5.8	20'
Kap Her	Marfik	Her	MS	16h 41m	+31° 36'	5.0	28"
Alp Her	Rasalgethi	Her	MS	17h 15m	+14° 23'	3.1	5"
Del Her	Sarin	Her	MS	17h 15m	+24° 50'	3.1	14"
Rho Her		Her	MS	17h 24m	+37° 09'	4.2	4"
95 Her		Her	MS	18h 02m	+21° 36'	4.3	6"
100 Her		Her	MS	18h 08m	+26° 06'	5.8	14"
Eps Lyr	The Double Double	Lyr	MS	18h 44m	+39° 40'	4.7	3"
Bet Lyr	Sheliak	Lyr	MS	18h 50m	+33° 22'	3.2	86"
Del Lyr		Lyr	MS	18h 54m	+36° 58'	4.2	630"
Zet Lyr		Lyr	MS	18h 45m	+37° 36'	4.4	44"
IC 4665	Summer Beehive	Oph	OC	17h 46m	+05° 43'	5.3	70'
M 12		Oph	GC	16h 47m	-01° 57'	6.1	16'
Rho Oph		Oph	MS	16h 26m	-23° 27'	4.6	3"
36 Oph		Oph	MS	17h 15m	-26° 36'	4.3	730"
Omi Oph		Oph	MS	17h 18m	-24° 17'	5.1	10"
61 Oph		Oph	MS	17h 45m	+02° 35'	6.2	21"
NGC 6633	Tweedledum Cluster	Oph	OC	18h 27m	+06° 31'	5.6	20'
70 Oph		Oph	MS	18h 06m	+02° 30'	4.2	4"
IC 4756		Ser	OC	18h 39m	+05° 27'	5.4	39'
M 16	Eagle Nebula	Ser	Neb	18h 19m	-13° 49'	6.0	9'
The Ser		Ser	MS	18h 56m	+04° 12'	4.3	22"
The Sge		Sge	MS	20h 10m	+20° 55'	4.6	84"
15 Sge		Sge	MS	20h 04m	+17° 04'	5.8	204"
M 17	Swan Nebula	Sgr	Neb	18h 21m	-16° 11'	6.0	11'
M 25		Sgr	OC	18h 32m	-19° 07'	6.2	29'
M 24	Sagittarius Star Cloud	Sgr	OC	18h 18m	-18° 24'	3.1	90'
M 22		Sgr	GC	18h 36m	-23° 54'	5.2	32'
M 8	Lagoon Nebula	Sgr	Neb	18h 04m	-24° 23'	5.0	17'
M 23		Sgr	OC	17h 57m	-18° 59'	5.9	29'
Collinder 399	Coathanger	Vul	Ast	19h 25m	+20° 11'	4.8	89'
NGC 6885		Vul	OC	20h 12m	+26° 29'	5.7	20'

Chart 20

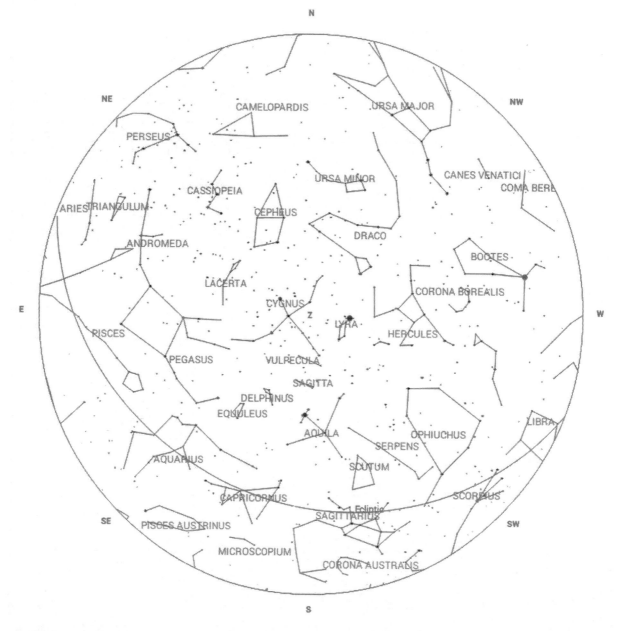

Designation	Name	Con.	Type	R.A.	Dec.	Mag	Size/Sep
15 Aql		Aql	MS	19h 06m	-04° 00'	5.4	39"
57 Aql		Aql	MS	19h 55m	-08° 14'	5.7	36"
Alp Cap	Al Giedi	Cap	MS	20h 18m	-12° 32'	3.6	45"
Bet Cap	Dabih	Cap	MS	20h 21m	-14° 47'	3.1	205"
Omi Cap		Cap	MS	20h 30m	-18° 35'	5.9	22"
IC 1396	Misty Clover Cluster	Cep	OC	21h 39m	+57° 30'	5.1	89'
Bet Cep	Alfirk	Cep	MS	21h 29m	+70° 34'	3.2	13"
Del Cep		Cep	MS/Var	22h 29m	+58° 25'	3.5-4.4	41"
Struve 2816		Cep	MS	21h 39m	+57° 29'	5.7	20"

Designation	Name	Con.	Type	R.A.	Dec.	Mag	Size/Sep
Mu Cep	Herschel's Garnet Star	Cep	Var/CS	21h 44m	+58° 47'	3.4-5.1	N/A
Struve 2840		Cep	MS	21h 52m	+55° 48'	5.7	18"
Xi Cep	Alkurhah	Cep	MS	22h 04m	+64° 38'	4.3	8"
Bet Cyg	Albireo	Cyg	MS	19h 31m	+27° 58'	3.1	34"
Del Cyg		Cyg	MS	19h 45m	+45° 08'	2.9	2"
16 Cyg		Cyg	MS	19h 42m	+50° 32'	5.9	39"
61 Cyg		Cyg	MS	21h 07m	+38° 44'	5.2	29"
M 39		Cyg	OC	21h 32m	+48° 26'	5.3	29'
Mu Cyg		Cyg	MS	21h 44m	+28° 45'	4.5	200"
NGC 7000	North American Nebula	Cyg	Neb	20h 59m	+44° 22'	4.0	120'
Gam Del		Del	MS	20h 47m	+16° 07'	3.9	9"
Nu Dra	Kuma	Dra	MS	17h 32m	+55° 10'	4.9	63"
Mu Dra		Dra	MS	17h 05m	+54° 28'	5.8	2"
Psi Dra		Dra	MS	17h 42m	+72° 09'	4.6	30"
39 Dra		Dra	MS	18h 24m	+58° 48'	5.0	89"
40/41 Dra		Dra	MS	18h 00m	+80° 00'	5.7	222"
Eps Dra		Dra	MS	19h 48m	+70° 16'	3.8	3"
Eps Equ		Equ	MS	20h 59m	+04° 18'	5.2	11"
Alp Her	Rasalgethi	Her	MS	17h 15m	+14° 23'	3.1	5"
Del Her	Sarin	Her	MS	17h 15m	+24° 50'	3.1	14"
Rho Her		Her	MS	17h 24m	+37° 09'	4.2	4"
95 Her		Her	MS	18h 02m	+21° 36'	4.3	6"
100 Her		Her	MS	18h 08m	+26° 06'	5.8	14"
8 Lac		Lac	MS	22h 36m	+39° 38'	5.7	82"
Eps Lyr	The Double Double	Lyr	MS	18h 44m	+39° 40'	4.7	3"
Bet Lyr	Sheliak	Lyr	MS	18h 50m	+33° 22'	3.2	86"
Del Lyr		Lyr	MS	18h 54m	+36° 58'	4.2	630"
Zet Lyr		Lyr	MS	18h 45m	+37° 36'	4.4	44"
IC 4665	Summer Beehive	Oph	OC	17h 46m	+05° 43'	5.3	70'
36 Oph		Oph	MS	17h 15m	-26° 36'	4.3	730"
Omi Oph		Oph	MS	17h 18m	-24° 17'	5.1	10"
NGC 6633	Tweedledum Cluster	Oph	OC	18h 27m	+06° 31'	5.6	20'
70 Oph		Oph	MS	18h 06m	+02° 30'	4.2	4"
Eps Peg	Enif	Peg	MS	21h 44m	+09° 52'	2.1	143"
IC 4756		Ser	OC	18h 39m	+05° 27'	5.4	39'
The Ser		Ser	MS	18h 56m	+04° 12'	4.3	22"
The Sge		Sge	MS	20h 10m	+20° 55'	4.6	84"
15 Sge		Sge	MS	20h 04m	+17° 04'	5.8	204"
M 24	Sagittarius Star Cloud	Sgr	OC	18h 18m	-18° 24'	3.1	90'
M 22		Sgr	GC	18h 36m	-23° 54'	5.2	32'
M 8	Lagoon Nebula	Sgr	Neb	18h 04m	-24° 23'	5.0	17'
M 23		Sgr	OC	17h 57m	-18° 59'	5.9	29'
Collinder 399	Coathanger	Vul	Ast	19h 25m	+20° 11'	4.8	89'
NGC 6885		Vul	OC	20h 12m	+26° 29'	5.7	20'

Chart 21

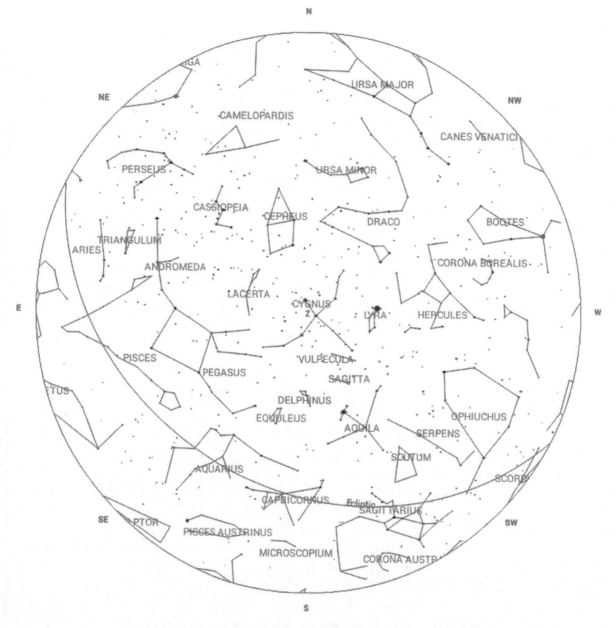

Designation	Name	Con.	Type	R.A.	Dec.	Mag	Size/Sep
Struve 2404		Aql	MS	18h 51m	+10° 59'	6.4	4"
15 Aql		Aql	MS	19h 06m	-04° 00'	5.4	39"
57 Aql		Aql	MS	19h 55m	-08° 14'	5.7	36"
V Aql		Aql	Var	19h 04m	-05° 41'	6.6-8.4	N/A
107 Aqr		Aqr	MS	23h 47m	-18° 35'	5.3	7"
94 Aqr		Aqr	MS	23h 19m	-13° 28'	5.2	13"
NGC 7293	Helix Nebula	Aqr	PN	22h 30m	-20° 50'	6.3	16'
Zet Aqr		Aqr	MS	22h 29m	-00° 01'	3.7	2"
41 Aqr		Aqr	MS	22h 14m	-21° 04'	5.3	5"

Designation	Name	Con.	Type	R.A.	Dec.	Mag	Size/Sep
53 Aqr		Aqr	MS	22h 27m	-16° 45'	5.6	3"
M 2		Aqr	GC	21h 33m	-00° 49'	6.6	16'
Alp Cap	Al Giedi	Cap	MS	20h 18m	-12° 32'	3.6	45"
Bet Cap	Dabih	Cap	MS	20h 21m	-14° 47'	3.1	205"
RT Cap		Cap	Var/CS	20h 17m	-21° 19'	6.5-8.1	N/A
Omi Cap		Cap	MS	20h 30m	-18° 35'	5.9	22"
Sig Cas		Cas	MS	23h 59m	+55° 45'	4.9	3"
IC 1396	Misty Clover Cluster	Cep	OC	21h 39m	+57° 30'	5.1	89'
Bet Cep	Alfirk	Cep	MS	21h 29m	+70° 34'	3.2	13"
Del Cep		Cep	MS/Var	22h 29m	+58° 25'	3.5-4.4	41"
Struve 2816		Cep	MS	21h 39m	+57° 29'	5.7	20"
Mu Cep	Herschel's Garnet Star	Cep	Var/CS	21h 44m	+58° 47'	3.4-5.1	N/A
Struve 2840		Cep	MS	21h 52m	+55° 48'	5.7	18"
Xi Cep	Alkurhah	Cep	MS	22h 04m	+64° 38'	4.3	8"
Omi Cep		Cep	MS	23h 19m	+68° 07'	4.8	3"
Bet Cyg	Albireo	Cyg	MS	19h 31m	+27° 58'	3.1	34"
Del Cyg		Cyg	MS	19h 45m	+45° 08'	2.9	2"
16 Cyg		Cyg	MS	19h 42m	+50° 32'	5.9	39"
61 Cyg		Cyg	MS	21h 07m	+38° 44'	5.2	29"
M 39		Cyg	OC	21h 32m	+48° 26'	5.3	29'
Mu Cyg		Cyg	MS	21h 44m	+28° 45'	4.5	200"
V460		Cyg	Var/CS	21h 42m	+35° 31'	5.6-7.0	N/A
NGC 7000	North American Nebula	Cyg	Neb	20h 59m	+44° 22'	4.0	120'
	Northern Coalsack	Cyg	DN	20h 41m	+43° 00'	6.0	60'
Gam Del		Del	MS	20h 47m	+16° 07'	3.9	9"
39 Dra		Dra	MS	18h 24m	+58° 48'	5.0	89"
40/41 Dra		Dra	MS	18h 00m	+80° 00'	5.7	222"
Eps Dra		Dra	MS	19h 48m	+70° 16'	3.8	3"
UX Dra		Dra	Var/CS	19h 22m	+76° 34'	5.9-7.1	N/A
Eps Equ		Equ	MS	20h 59m	+04° 18'	5.2	11"
95 Her		Her	MS	18h 02m	+21° 36'	4.3	6"
100 Her		Her	MS	18h 08m	+26° 06'	5.8	14"
8 Lac		Lac	MS	22h 36m	+39° 38'	5.7	82"
Eps Lyr	The Double Double	Lyr	MS	18h 44m	+39° 40'	4.7	3"
Bet Lyr	Sheliak	Lyr	MS	18h 50m	+33° 22'	3.2	86"
Del Lyr		Lyr	MS	18h 54m	+36° 58'	4.2	630"
Zet Lyr		Lyr	MS	18h 45m	+37° 36'	4.4	44"
M 15		Peg	GC	21h 30m	+12° 10'	6.3	18'
Eps Peg	Enif	Peg	MS	21h 44m	+09° 52'	2.1	143"
M 11	Wild Duck Cluster	Sct	OC	18h 51m	-06° 16'	6.1	32'
The Sge		Sge	MS	20h 10m	+20° 55'	4.6	84"
15 Sge		Sge	MS	20h 04m	+17° 04'	5.8	204"
Collinder 399	Coathanger	Vul	Ast	19h 25m	+20° 11'	4.8	89'
NGC 6885		Vul	OC	20h 12m	+26° 29'	5.7	20'

Chart 22

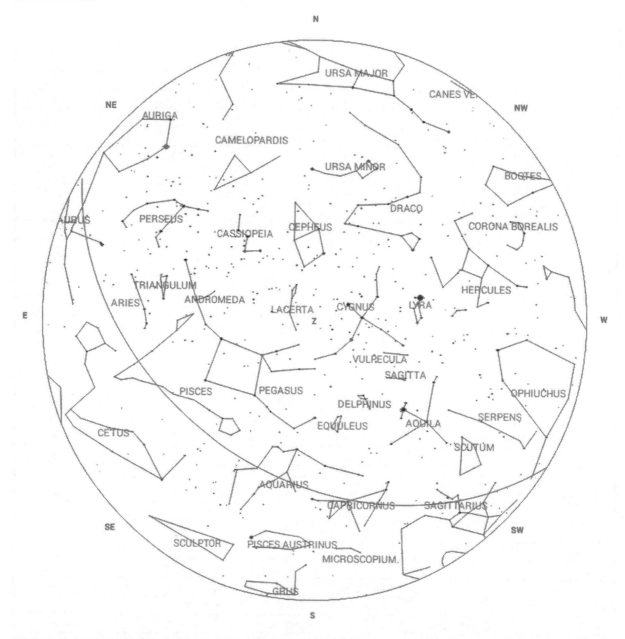

Designation	Name	Con.	Type	R.A.	Dec.	Mag	Size/Sep
M 31	Andromeda Galaxy	And	Gx	00h 43m	+41° 16'	4.3	156'
Pi And		And	MS	00h 38m	+33° 49'	4.4	36"
15 Aql		Aql	MS	19h 06m	-04° 00'	5.4	39"
57 Aql		Aql	MS	19h 55m	-08° 14'	5.7	36"
107 Aqr		Aqr	MS	23h 47m	-18° 35'	5.3	7"
94 Aqr		Aqr	MS	23h 19m	-13° 28'	5.2	13"
NGC 7293	Helix Nebula	Aqr	PN	22h 30m	-20° 50'	6.3	16'
Zet Aqr		Aqr	MS	22h 29m	-00° 01'	3.7	2"

Designation	Name	Con.	Type	R.A.	Dec.	Mag	Size/Sep
41 Aqr		Aqr	MS	22h 14m	-21° 04'	5.3	5"
53 Aqr		Aqr	MS	22h 27m	-16° 45'	5.6	3"
M 2		Aqr	GC	21h 33m	-00° 49'	6.6	16'
M 30		Cap	GC	21h 40m	-23° 11'	6.9	12'
Alp Cap	Al Giedi	Cap	MS	20h 18m	-12° 32'	3.6	45"
Bet Cap	Dabih	Cap	MS	20h 21m	-14° 47'	3.1	205"
Omi Cap		Cap	MS	20h 30m	-18° 35'	5.9	22"
Struve 3053		Cas	MS	00h 03m	+66° 06'	5.9	15"
Eta Cas	Achird	Cas	MS	00h 50m	+57° 54'	3.6	13"
Sig Cas		Cas	MS	23h 59m	+55° 45'	4.9	3"
IC 1396	Misty Clover Cluster	Cep	OC	21h 39m	+57° 30'	5.1	89'
Bet Cep	Alfirk	Cep	MS	21h 29m	+70° 34'	3.2	13"
Del Cep		Cep	MS/Var	22h 29m	+58° 25'	3.5-4.4	41"
Struve 2816		Cep	MS	21h 39m	+57° 29'	5.7	20"
Mu Cep	Herschel's Garnet Star	Cep	Var/CS	21h 44m	+58° 47'	3.4-5.1	N/A
Struve 2840		Cep	MS	21h 52m	+55° 48'	5.7	18"
Xi Cep	Alkurhah	Cep	MS	22h 04m	+64° 38'	4.3	8"
Omi Cep		Cep	MS	23h 19m	+68° 07'	4.8	3"
Bet Cyg	Albireo	Cyg	MS	19h 31m	+27° 58'	3.1	34"
Del Cyg		Cyg	MS	19h 45m	+45° 08'	2.9	2"
16 Cyg		Cyg	MS	19h 42m	+50° 32'	5.9	39"
NGC 6960	Veil Nebula (West)	Cyg	SNR	20h 46m	+30° 43'	7.0	63'
61 Cyg		Cyg	MS	21h 07m	+38° 44'	5.2	29"
M 39		Cyg	OC	21h 32m	+48° 26'	5.3	29'
Mu Cyg		Cyg	MS	21h 44m	+28° 45'	4.5	200"
V460		Cyg	Var/CS	21h 42m	+35° 31'	5.6-7.0	N/A
NGC 6992	Veil Nebula (East)	Cyg	SNR	20h 56m	+31° 43'	7.0	60'
NGC 7000	North American Nebula	Cyg	Neb	20h 59m	+44° 22'	4.0	120'
	Northern Coalsack	Cyg	DN	20h 41m	+43° 00'	6.0	60'
Gam Del		Del	MS	20h 47m	+16° 07'	3.9	9"
Eps Dra		Dra	MS	19h 48m	+70° 16'	3.8	3"
Eps Equ		Equ	MS	20h 59m	+04° 18'	5.2	11"
NGC 7243		Lac	OC	22h 15m	+49° 54'	6.7	29'
8 Lac		Lac	MS	22h 36m	+39° 38'	5.7	82"
Struve 2470/2474	The Double Double's Double	Lyr	MS	19h 09m	+34° 41'	6.7	16"
M 15		Peg	GC	21h 30m	+12° 10'	6.3	18'
Eps Peg	Enif	Peg	MS	21h 44m	+09° 52'	2.1	143"
TX Psc		Psc	Var/CS	23h 46m	+03° 29'	4.5-5.3	N/A
55 Psc		Psc	MS	00h 40m	+21° 26'	5.4	6"
65 Psc		Psc	MS	00h 50m	+27° 43'	7.0	4"
The Sge		Sge	MS	20h 10m	+20° 55'	4.6	84"
15 Sge		Sge	MS	20h 04m	+17° 04'	5.8	204"
Collinder 399	Coathanger	Vul	Ast	19h 25m	+20° 11'	4.8	89'
NGC 6885		Vul	OC	20h 12m	+26° 29'	5.7	20'

Chart 23

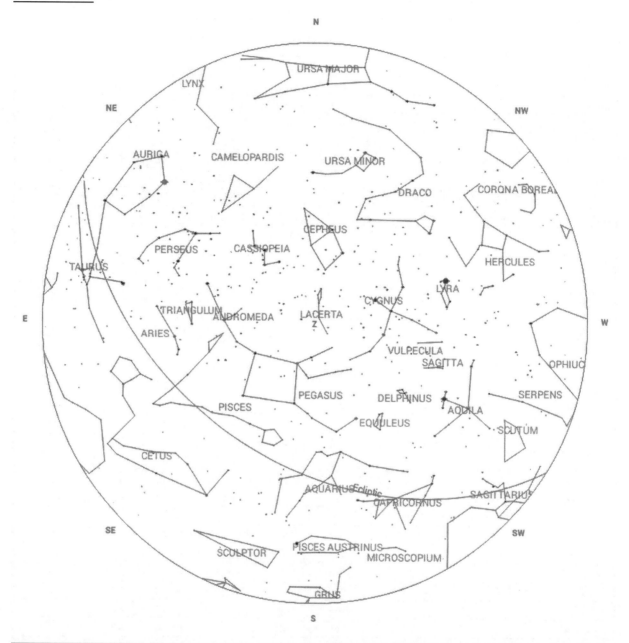

Designation	Name	Con.	Type	R.A.	Dec.	Mag	Size/Sep
M 31	Andromeda Galaxy	And	Gx	00h 43m	+41° 16'	4.3	156'
NGC 752	Golf Ball Cluster	And	OC	01h 58m	+37° 47'	6.6	75'
Pi And		And	MS	00h 38m	+33° 49'	4.4	36"
107 Aqr		Aqr	MS	23h 47m	-18° 35'	5.3	7"
94 Aqr		Aqr	MS	23h 19m	-13° 28'	5.2	13"
NGC 7293	Helix Nebula	Aqr	PN	22h 30m	-20° 50'	6.3	16'
Zet Aqr		Aqr	MS	22h 29m	-00° 01'	3.7	2"
41 Aqr		Aqr	MS	22h 14m	-21° 04'	5.3	5"

Designation	Name	Con.	Type	R.A.	Dec.	Mag	Size/Sep
53 Aqr		Aqr	MS	22h 27m	-16° 45'	5.6	3"
M 2		Aqr	GC	21h 33m	-00° 49'	6.6	16'
Lam Ari		Ari	MS	01h 59m	+23° 41'	4.8	37"
Gam Ari	Mesarthim	Ari	MS	01h 54m	+19° 22'	4.6	8"
M 30		Cap	GC	21h 40m	-23° 11'	6.9	12'
Alp Cap	Al Giedi	Cap	MS	20h 18m	-12° 32'	3.6	45"
Bet Cap	Dabih	Cap	MS	20h 21m	-14° 47'	3.1	205"
Omi Cap		Cap	MS	20h 30m	-18° 35'	5.9	22"
NGC 457	Owl Cluster	Cas	OC	01h 20m	+58° 17'	5.1	20'
Struve 163		Cas	MS	01h 51m	+64° 51'	6.5	35"
M 103		Cas	OC	01h 33m	+60° 39'	6.9	5'
Struve 3053		Cas	MS	00h 03m	+66° 06'	5.9	15"
Eta Cas	Achird	Cas	MS	00h 50m	+57° 54'	3.6	13"
Sig Cas		Cas	MS	23h 59m	+55° 45'	4.9	3"
NGC 663		Cas	OC	01h 46m	+61° 14'	6.4	14'
IC 1396	Misty Clover Cluster	Cep	OC	21h 39m	+57° 30'	5.1	89'
Bet Cep	Alfirk	Cep	MS	21h 29m	+70° 34'	3.2	13"
Del Cep		Cep	MS/Var	22h 29m	+58° 25'	3.5-4.4	41"
Struve 2816		Cep	MS	21h 39m	+57° 29'	5.7	20"
Mu Cep	Herschel's Garnet Star	Cep	Var/CS	21h 44m	+58° 47'	3.4-5.1	N/A
Struve 2840		Cep	MS	21h 52m	+55° 48'	5.7	18"
Xi Cep	Alkurhah	Cep	MS	22h 04m	+64° 38'	4.3	8"
Omi Cep		Cep	MS	23h 19m	+68° 07'	4.8	3"
NGC 6960	Veil Nebula (West)	Cyg	SNR	20h 46m	+30° 43'	7.0	63'
61 Cyg		Cyg	MS	21h 07m	+38° 44'	5.2	29"
M 39		Cyg	OC	21h 32m	+48° 26'	5.3	29'
Mu Cyg		Cyg	MS	21h 44m	+28° 45'	4.5	200"
NGC 6992	Veil Nebula (East)	Cyg	SNR	20h 56m	+31° 43'	7.0	60'
NGC 7000	North American Nebula	Cyg	Neb	20h 59m	+44° 22'	4.0	120'
	Northern Coalsack	Cyg	DN	20h 41m	+43° 00'	6.0	60'
Gam Del		Del	MS	20h 47m	+16° 07'	3.9	9"
Eps Equ		Equ	MS	20h 59m	+04° 18'	5.2	11"
NGC 7243		Lac	OC	22h 15m	+49° 54'	6.7	29'
8 Lac		Lac	MS	22h 36m	+39° 38'	5.7	82"
M 15		Peg	GC	21h 30m	+12° 10'	6.3	18'
Eps Peg	Enif	Peg	MS	21h 44m	+09° 52'	2.1	143"
55 Psc		Psc	MS	00h 40m	+21° 26'	5.4	6"
65 Psc		Psc	MS	00h 50m	+27° 43'	7.0	4"
Psi 1 Psc		Psc	MS	01h 06m	+21° 28'	5.3	30"
Zet Psc		Psc	MS	01h 14m	+07° 35'	5.2	23"
The Sge		Sge	MS	20h 10m	+20° 55'	4.6	84"
15 Sge		Sge	MS	20h 04m	+17° 04'	5.8	204"
M 33	Triangulum Galaxy	Tri	Gx	01h 34m	+30° 40'	6.4	62'
NGC 6885		Vul	OC	20h 12m	+26° 29'	5.7	20'

Chart 24

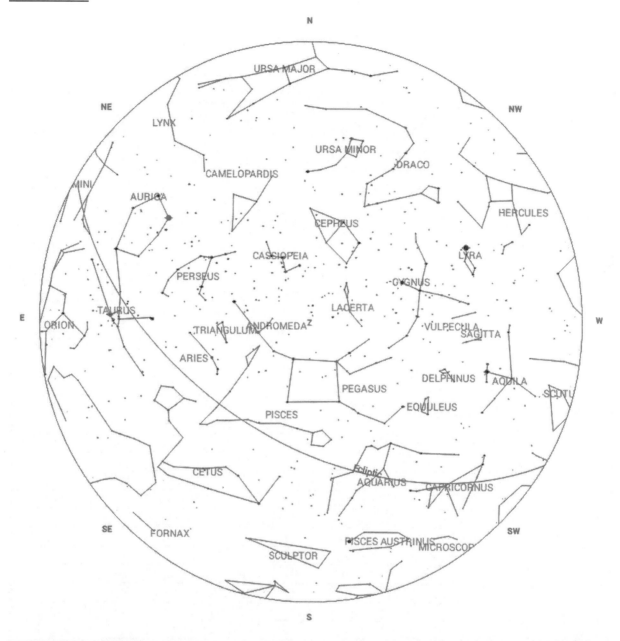

Designation	Name	Con.	Type	R.A.	Dec.	Mag	Size/Sep
Gam And	Almach	And	MS	02h 04m	+42° 20'	2.1	10"
M 31	Andromeda Galaxy	And	Gx	00h 43m	+41° 16'	4.3	156'
NGC 752	Golf Ball Cluster	And	OC	01h 58m	+37° 47'	6.6	75'
Pi And		And	MS	00h 38m	+33° 49'	4.4	36"
107 Aqr		Aqr	MS	23h 47m	-18° 35'	5.3	7"
94 Aqr		Aqr	MS	23h 19m	-13° 28'	5.2	13"
NGC 7293	Helix Nebula	Aqr	PN	22h 30m	-20° 50'	6.3	16'
Zet Aqr		Aqr	MS	22h 29m	-00° 01'	3.7	2"

Designation	Name	Con.	Type	R.A.	Dec.	Mag	Size/Sep
41 Aqr		Aqr	MS	22h 14m	-21° 04'	5.3	5"
53 Aqr		Aqr	MS	22h 27m	-16° 45'	5.6	3"
M 2		Aqr	GC	21h 33m	-00° 49'	6.6	16'
Lam Ari		Ari	MS	01h 59m	+23° 41'	4.8	37"
Gam Ari	Mesarthim	Ari	MS	01h 54m	+19° 22'	4.6	8"
30 Ari		Ari	MS	02h 37m	+24° 39'	6.5	39"
M 30		Cap	GC	21h 40m	-23° 11'	6.9	12'
NGC 457	Owl Cluster	Cas	OC	01h 20m	+58° 17'	5.1	20'
Struve 163		Cas	MS	01h 51m	+64° 51'	6.5	35"
M 103		Cas	OC	01h 33m	+60° 39'	6.9	5'
Struve 3053		Cas	MS	00h 03m	+66° 06'	5.9	15"
Eta Cas	Achird	Cas	MS	00h 50m	+57° 54'	3.6	13"
Sig Cas		Cas	MS	23h 59m	+55° 45'	4.9	3"
Iot Cas		Cas	MS	02h 29m	+67° 24'	4.5	7"
NGC 659	Ying Yang Cluster	Cas	OC	01h 44m	+60° 40'	7.2	5'
NGC 663		Cas	OC	01h 46m	+61° 14'	6.4	14'
IC 1396	Misty Clover Cluster	Cep	OC	21h 39m	+57° 30'	5.1	89'
Bet Cep	Alfirk	Cep	MS	21h 29m	+70° 34'	3.2	13"
Del Cep		Cep	MS/Var	22h 29m	+58° 25'	3.5-4.4	41"
Struve 2816		Cep	MS	21h 39m	+57° 29'	5.7	20"
Mu Cep	Herschel's Garnet Star	Cep	Var/CS	21h 44m	+58° 47'	3.4-5.1	N/A
Struve 2840		Cep	MS	21h 52m	+55° 48'	5.7	18"
Xi Cep	Alkurhah	Cep	MS	22h 04m	+64° 38'	4.3	8"
Omi Cep		Cep	MS	23h 19m	+68° 07'	4.8	3"
Gam Cet	Kaffajidhma	Cet	MS	02h 43m	+03° 14'	3.5	3"
Omi Cet	Mira	Cet	Var	02h 19m	-02° 59'	2.0-10.1	N/A
61 Cyg		Cyg	MS	21h 07m	+38° 44'	5.2	29"
M 39		Cyg	OC	21h 32m	+48° 26'	5.3	29'
Mu Cyg		Cyg	MS	21h 44m	+28° 45'	4.5	200"
V460		Cyg	Var/CS	21h 42m	+35° 31'	5.6-7.0	N/A
NGC 7243		Lac	OC	22h 15m	+49° 54'	6.7	29'
8 Lac		Lac	MS	22h 36m	+39° 38'	5.7	82"
M 34		Per	OC	02h 42m	+42° 46'	5.8	35'
NGC 869/884	Double Cluster	Per	OC	02h 21m	+57° 08'	4.4	18'
Eta Per		Per	MS	02h 51m	+55° 54'	3.8	29"
M 15		Peg	GC	21h 30m	+12° 10'	6.3	18'
Eps Peg	Enif	Peg	MS	21h 44m	+09° 52'	2.1	143"
Alp Psc	Alrisha	Psc	MS	02h 02m	+02° 46'	3.8	2"
TX Psc		Psc	Var/CS	23h 46m	+03° 29'	4.5-5.3	N/A
55 Psc		Psc	MS	00h 40m	+21° 26'	5.4	6"
Psi1 Psc		Psc	MS	01h 06m	+21° 28'	5.3	30"
Zet Psc		Psc	MS	01h 14m	+07° 35'	5.2	23"
M 33	Triangulum Galaxy	Tri	Gx	01h 34m	+30° 40'	6.4	62'
Alp UMi	Polaris	UMi	MS	02h 51m	+89° 20'	2.0	18"

Visibility Table

This table summarizes which sights can be seen on a month-by-month basis throughout the year. The following icons are used:

●	New Moon	◑	Waxing Crescent	◐	First Quarter	◑	Waxing Gibbous
○	Full Moon	◐	Waning Gibbous	◑	Last Quarter	◐	Waning Crescent
☿	Mercury	♀	Venus	♂	Mars	♃	Jupiter
♄	Saturn	⛢	Uranus	♆	Neptune		
****	Star Cluster	*	Bright Star				

For each ten day period the Moon and planets are listed as being Not Visible (NV), Evening Sky (PM), All Night (AN) or Morning Sky (AM) as applicable. If an object appears close to another object during this time then the other object is listed on the line below. For example, between January 1st and January 10th the waning crescent Moon appears close to Mars and Jupiter.

N.B. This table is meant to be a rough, at-a-glance guide. The Moon and Mercury move quickly and it's worth bearing in mind that although these worlds may appear close to a planet or bright star at some point during that ten day period, they may not appear close to that planet or bright star for the *entire* ten day period. Please check the relevant Daily Details for more precise information. Likewise, the Moon may be waning, new and then waxing again within the same timeframe (e.g., January 21st to 31st.)

January

1st – 10th

○	◑	◐	●	☿	♀	♂	♃	♄	⛢	♆
AN	AM	AM	AM	AM	NV	AM	AM	AM	PM	PM
*	*	*	♂ ♃	♄		● ♃	● ♂	☿		

11th – 20th

◐	●			☿	♀	♂	♃	♄	⛢	♆
AM	NV			AM	NV	AM	AM	AM	PM	PM
♂ ♃	☿ ♀ ♄ ♆			● ♄	●	◐ ♃	● ♂	● ☿		●

21st – 31st

●	◐	◑	○	☿	♀	♂	♃	♄	⛢	♆
PM	PM	PM	AN	NV	NV	AM	AM	AM	PM	PM
	⛢	**** *	****			♃ *	♂		◑	

Month	Period											
February	1st – 10th	○ AN *	◑ AM *	◑ AM ♂ ♃ *	◑ AM ♂ *	♇ NV	♀ NV	♂ AM ◑ ◑ *	♃ AM ◑	♄ AM	⛢ PM	♆ PM
February	11th – 20th	◐ AM ♄	● NV ♇ ♀ ♆	◐ PM ⛢		♇ NV ● ♀	♀ NV ● ♇ ♆	♂ AM *	♃ AM	♄ AM ◐	⛢ PM ◐	♆ NV ●
February	21st – 28th	◐ PM	◐ PM **** *	◐ PM ****	○ AN **** *	♇ NV ♀ ♆	♀ NV ♇ ♆	♂ AM *	♃ AM	♄ AM	⛢ PM	♆ NV ♇ ♀
March	1st – 10th	○ AN	◑ AM ♃ *	◑ AM ♂ ♄ *		♇ NV ♀	♀ NV ♇	♂ AM ◑	♃ AM ◑	♄ AM ◑	⛢ PM	♆ NV
March	11th – 20th	◐ AM ♂ ♄	● NV ♇ ♀ ⛢ ♆			♇ PM ● ♀	♀ PM ● ♇	♂ AM ◐ ♄	♃ AM	♄ AM ◐ ♂	⛢ PM ●	♆ NV ●
March	21st – 31st	◐ PM **** *	◐ PM	◐ PM **** *	○ AN *	♇ NV ♀	♀ PM ♇ ⛢	♂ AM ♄	♃ AM	♄ AM ♂	⛢ PM ♀	♆ NV

April	1st – 10th	○ AN *	◑ AM ♃ *	◑ AM ♂ ♄	◕ AM	☿ NV	♀ PM ⛢	♂ AM ◑ ♄	♃ AM ◑	♄ AM ◑ ♂	⛢ NV ♀	♆ AM
	11th – 20th	◕ AM ♆	● NV ☿ ♀ ⛢ ****	● PM **** *		☿ AM ●	♀ PM ● ****	♂ AM ♄	♃ AM	♄ AM	⛢ NV ● ☿ ♀	♆ AM ◕
	21st – 30th	◖ PM	◐ PM ****	◐ PM *	○ AN ♃ *	☿ AM	♀ NV **** *	♂ AM	♃ AM ○	♄ AM	⛢ NV	♆ AM
May	1st – 10th	○ AN ♃ *	◑ AM ♂ ♄ *	◑ AM	◕ AM ♆	☿ AM ⛢	♀ PM **** *	♂ AM ◑	♃ AN ○	♄ AM ◑	⛢ AM	♆ AM
	11th – 20th	◕ AM	● NV ☿ ♀ ⛢ **** *	● PM ****		☿ AM ● ⛢	♀ PM ●	♂ AM	♃ AN	♄ AM	⛢ AM ● ☿	♆ AM
	21st – 31tt	◐ PM *	◐ PM *	○ AN ♃ ♄ *		☿ NV **** *	♀ PM	♂ AM	♃ PM ○	♄ AM ○	⛢ AM	♆ AM

Month	Period											
June	1st – 10th	◑ AM	◑ AM	◕ AM		♇ NV	♀ PM	♂ AM	♃ PM	♄ AM	⛢ AM	♆ AM
		♂ ♄	♆	⛢		**** *	◑			◑	◕	◑
	11th – 20th	● NV	◕ PM	◐ PM		♇ NV	♀ PM	♂ AM	♃ PM	♄ AM	⛢ AM	♆ AM
		♇ ♀ **** *	**** *			●	● ****					
	21st – 30th	◐ PM	◐ PM	○ AN	◑ AM	♇ PM	♀ PM	♂ AM	♃ PM	♄ AN	⛢ AM	♆ AM
		*	♃ *	♄ *	♂		****	◑	◐	○		
July	1st – 10th	◑ AM	◑ AM	◕ AM	● NV	♇ PM	♀ PM	♂ AM	♃ PM	♄ PM	⛢ AM	♆ AM
		♂ ♆	⛢	****	**** *	****	*	◑				◑
	11th – 20th	● NV	◕ PM	◐ PM		♇ PM	♀ PM	♂ AM	♃ PM	♄ PM	⛢ AM	♆ AM
		♇ ♀ **** *	♀ *	♃ *		● *	● ◕ *		◐			
	21st – 31tt	◐ PM	○ AN	◑ AM		♇ PM	♀ PM	♂ AN	♃ PM	♄ PM	⛢ AM	♆ AM
		♃ ♄ *	♂ ♄	♆		*		○	◐	◐ ○		◑

Month	Dates											
August	1st – 10th	🌖 AM	◑ AM ⛢ ****	🌘 AM **** *	● NV ♇ ****	♇ NV ● *	♀ PM	♂ PM	♃ PM	♄ PM	⛢ AM ◑	♆ AM
	11th – 20th	● NV ♇ *	🌒 PM ♀ ♃ *	◐ PM ♃ *	🌔 PM ♄	♇ NV ● ****	♀ PM 🌒	♂ PM	♃ PM 🌒 ◐	♄ PM 🌔	⛢ AM	♆ AM
	21st – 31st	🌔 PM ♄ ♂	○ AN ♂ ♆	🌖 AM ⛢		♇ AM **** *	♀ PM *	♂ PM 🌔 ○	♃ PM	♄ AN 🌔	⛢ AM 🌖	♆ AM ○
September	1st – 10th	🌖 AM ****	◑ AM **** *	🌘 AM ****	● NV ♇ **** *	♇ NV ● *	♀ PM *	♂ PM	♃ PM	♄ PM	⛢ AM	♆ AN
	11th – 20th	● NV ♀ *	🌒 PM ♀ ♃ *	◐ PM ♄ *	🌔 PM ♂	♇ NV	♀ PM ● 🌒 *	♂ PM 🌔	♃ PM 🌒	♄ PM ◐	⛢ AM	♆ AN
	21st – 30th	🌔 PM	○ AN ♆	🌖 AM ⛢ **** *		♇ NV *	♀ PM	♂ PM	♃ PM	♄ PM	⛢ AM	♆ PM

Month	Dates											
October	1st – 10th	◑ AM	◕ AM **** *	● NV ♇ ♀ *		♇ NV ● ♀ *	♀ PM ● ♇	♂ PM	♃ PM	♄ PM	♅ AM	♆ PM
October	11th – 20th	● NV ♀ ♃	◕ PM ♃ ♄ *	◐ PM ♂ ♄	◔ PM ♂ ♆	♇ PM ♀ *	♀ PM ● ♇	♂ PM ◐ ◔	♃ PM ● ◕	♄ PM ◕ ◐	♅ AN	♆ PM ◔
October	21st – 31st	◔ PM	○ AN ♅ ****	◑ AM **** *	◑ AM ****	♇ PM ♃ *	♀ NV *	♂ PM	♃ PM ♇	♄ PM	♅ AN ○	♆ PM
November	1st – 10th	◑ AM **** *	◕ AM *	● NV ♇ ♀ ♃ *		♇ PM ● ♃ *	♀ AM ● *	♂ PM	♃ PM ● ♇ *	♄ PM	♅ PM	♆ PM
November	11th – 20th	◕ PM ♄	◐ PM ♂	◔ PM ♅ ♆		♇ PM *	♀ AM *	♂ PM ◐	♃ NV *	♄ PM ◕	♅ PM ◔	♆ PM ◔
November	21st – 30th	○ AN **** *	◑ AM ****	◑ AM *		♇ NV ♃ *	♀ AM *	♂ PM ♆	♃ NV ♇ *	♄ PM	♅ PM	♆ PM ♂

December

1st – 10th

◑ AM	● NV			☿ AM	♀ AM	♂ PM	♃ NV	♄ PM	♅ PM	♆ PM
♀ *	☿ ♀ ♃ ♄ *			● ♃ *	◑ ● *	♆ ☿	●	●		♂

11th – 20th

◐ PM	◐ PM	◑ PM		☿ AM	♀ AM	♂ PM	♃ NV	♄ PM	♅ PM	♆ PM
♂ ♆	♅ **** *			♃ *		◐ ♆	☿ *			◑ ♂

21st – 31st

○ AN	◐ AM	◑ AM	● AM	☿ AM	♀ AM	♂ PM	♃ AM	♄ NV	♅ PM	♆ PM
**** *	**** *	*		♃ *		☿ *				

Meteor Showers for 2018

In the table below, the Zenith Hourly Rate (ZHR) indicates how many meteors you might expect to see every hour under ideal situations.

Active Dates	Peak Date	Shower Name	Zenith Hourly Rate (ZHR)	Moon Phase
December 28th to January 12th	January 4th	Quadrantids	120	◑
April 16th to April 25th	April 22nd	Lyrids	18	◐
April 19th to May 28th	May 5th	Eta Aquariids	65	○
May 22nd to July 2nd	June 7th	Arietids	54	○
July 12th to August 23rd	July 29th	Delta Aquariids	16	○
July 17th to August 24th	August 12th	Perseids	100	●
October 6th to October 10th	October 8th	Draconids	Var.	●
October 2nd to November 7th	October 21st	Orionids	25	◑
November 6th to November 30th	November 17th	Leonids	15	◐
December 4th to December 17th	December 13th	Geminids	120	◐
December 17th to December 26th	December 23rd	Ursids	10	○

Key to the Monthly Guide Lunar and Planetary Data Tables

The table below provides a key to the abbreviations and acronyms used in the data tables for the Moon and planets. Please note that the position data (ie, the R.A. and declination) are accurate for 12:00 Universal Time.

+Cr	Waxing Crescent Moon	FQ	First Quarter Moon	LQ	Last Quarter Moon
-Cr	Waning Crescent Moon	+G	Waxing Gibbous Moon	NM	New Moon
FM	Full Moon	-G	Waning Gibbous Moon		
AM	Morning visibility	Gem	Gemini	PM	Evening Sky
Aqr	Aquarius	Ill	Illumination	Psc	Pisces
Ari	Aries	Leo	Leo	R.A.	Right Ascension
Cap	Capricornus	Lib	Libra	Sco	Scorpius
Cet	Cetus	Mag	Magnitude	Sex	Sextans
Con	Constellation	NV	Not Visible	Sgr	Sagittarius
Dec	Declination	Oph	Ophiuchus	Tau	Taurus
Diam	Apparent Diameter	Ori	Orion	Vir	Virgo
Elong	Elongation from Sun				

About the Events

All of the times listed are in Universal Time. I considered changing the times to Eastern Time but then thought it would still be inconvenient for those located in other areas of the country. With that in mind, I've included a conversion table below. Also, bear in mind that if an event is not visible at the specific time it occurs (e.g., the Moon appearing close to a planet) it will always sbe worth observing at the nearest convenient opportunity.

You'll also notice that I don't provide angular separation for conjunctions involving the Moon. Since the Moon is so much closer than the planets, the angular separation will vary depending upon your latitude upon the Earth. With that in mind, I've stuck to stating that the Moon simply appears *close* to a particular planet, star or star cluster in the sky.

	Standard Time	Summer Time
Eastern Time	UT-5 hours	UT-4 hours
Central Time	UT-6 hours	UT-5 hours
Mountain Time	UT-7 hours	UT-6 hours
Pacific Time	UT-8 hours	UT-7 hours

The Moon

| 1st | 3rd | 5th | 7th | 9th |

Date	Con	R.A.	Dec	Mag	Diam	Ill.	Elong	Phase	Close To
1st	Ori	06h 10m	19° 51'	-12.4	33'	99%	171° E	FM	
2nd	Gem	07h 16m	19° 51'	-12.5	33'	100%	174° W	FM	
3rd	Cnc	08h 20m	18° 23'	-12.1	33'	97%	160° W	FM	Praesepe
4th	Cnc	09h 21m	15° 42'	-11.6	33'	92%	146° W	-G	Praesepe, Regulus
5th	Leo	10h 19m	12° 5'	-11.2	32'	84%	133° W	-G	Regulus
6th	Leo	11h 13m	7° 54'	-10.8	32'	75%	120° W	-G	
7th	Vir	12h 05m	3° 27'	-10.4	31'	65%	107° W	LQ	
8th	Vir	12h 55m	-1° 3'	-10.0	31'	55%	95° W	LQ	Spica
9th	Vir	13h 43m	-5° 21'	-9.5	31'	44%	83° W	LQ	Spica
10th	Lib	14h 30m	-9° 18'	-9.1	30'	35%	72° W	-Cr	Mars, Jupiter

Mercury and Venus

| Mercury | Mercury | Venus |
| 3rd | 7th | 5th |

Mercury

Date	Con	R.A.	Dec	Mag	Diam	Ill.	Elong	AM/PM	Close To
1st	Oph	17h 10m	-21° 0'	-0.3	7''	63%	23° W	AM	
3rd	Oph	17h 19m	-21° 27'	-0.3	6''	68%	23° W	AM	
5th	Oph	17h 29m	-21° 54'	-0.3	6''	71%	22° W	AM	Saturn
7th	Oph	17h 39m	-22° 18'	-0.3	6''	75%	22° W	AM	Saturn
9th	Sgr	17h 50m	-22° 39'	-0.3	6''	78%	21° W	AM	Saturn

Venus

Date	Con	R.A.	Dec	Mag	Diam	Ill	Elong	AM/PM	Close To
1st	Sgr	18h 40m	-23° 36'	-3.9	10''	100%	2° W	NV	Saturn
3rd	Sgr	18h 51m	-23° 30'	-3.9	10''	100%	2° W	NV	
5th	Sgr	19h 02m	-23° 18'	-3.9	10''	100%	1° W	NV	
7th	Sgr	19h 13m	-23° 6'	-3.9	10''	100%	1° W	NV	
9th	Sgr	19h 24m	-22° 48'	-3.9	10''	100%	1° E	NV	

Mars and the Outer Planets

Mars
5th

Jupiter
5th

Saturn
5th

Mars

Date	Con	R.A.	Dec	Mag	Diam	Ill	Elong	AM/PM	Close To
1st	Lib	14h 49m	-15° 18'	1.5	5''	93%	57° W	AM	Jupiter
5th	Lib	14h 59m	-16° 3'	1.4	5''	93%	58° W	AM	Jupiter
10th	Lib	15h 12m	-16° 57'	1.4	5''	93%	60° W	AM	Moon, Jupiter

The Outer Planets

Planet	Date	Con	R.A.	Dec	Mag	Diam	Elong	AM/PM	Close To
Jupiter	5th	Lib	15h 02m	-16° 6'	-1.8	33''	57° W	AM	Mars
Saturn	5th	Sgr	18h 08m	-22° 30'	0.5	15''	13° W	NV	Mercury
Uranus	5th	Psc	01h 32m	-9° 0'	5.8	4''	100° E	PM	
Neptune	5th	Aqr	22h 55m	-7° 54'	7.9	2''	57° E	PM	

Events

Date	Time (UT)	Event
1st	19:46	Mercury is at greatest western elongation. (Morning sky.)
2nd	02:25	Full Moon. (All night.)
	19:13	Uranus is stationary prior to resuming prograde motion. (Evening sky.)
3rd	07:12	The Earth is at its minimum distance from the Sun
4th	N/A	The Quadrantid meteors are at their peak. (ZHR: 120. All night but best in the morning sky.)
5th	09:09	The waning gibbous Moon appears north of the bright star Regulus. (Leo, morning sky.)
6th	09:35	Mercury is at its brightest. (Morning sky.)
7th	03:37	Mars appears 0.2° south of Jupiter. (Morning sky.)
8th	22:26	Last Quarter Moon. (Morning sky.)
9th	01:17	The just-past last quarter Moon appears north of the bright star Spica. (Virgo, morning sky.)
	02:14	Dwarf planet Pluto is in conjunction with the Sun. (Not visible.)
	06:21	Venus is at superior conjunction with the Sun. (Not visible.)

Planet Locations – January 5th

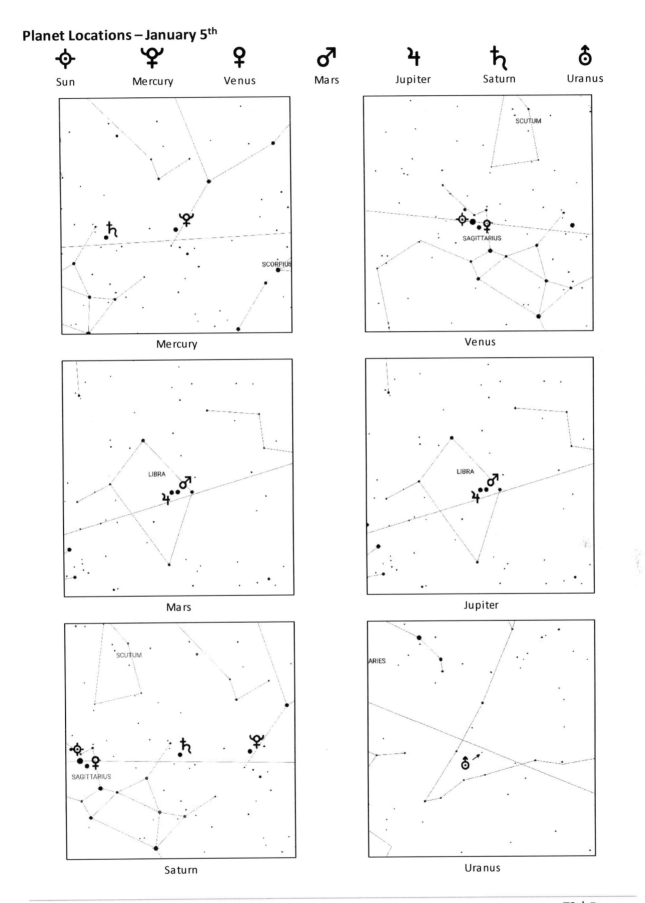

January 11th – 20th

The Moon

| 11th | 13th | 15th | 17th | 19th |

Date	Con	R.A.	Dec	Mag	Diam	Ill.	Elong	Phase		Close To
11th	Lib	15h 18m	-12° 48'	-8.5	30'	26%	61° W	-Cr		Mars, Jupiter
12th	Sco	16h 06m	-15° 45'	-7.9	30'	18%	50° W	-Cr		Mars, Antares
13th	Oph	16h 54m	-17° 57'	-7.3	30'	11%	39° W	NM		Antares
14th	Oph	17h 44m	-19° 24'	-6.6	30'	6%	28° W	NM		Mercury, Saturn
15th	Sgr	18h 34m	-20° 3'	-5.8	30'	2%	18° W	NM		Mercury, Saturn
16th	Sgr	19h 24m	-19° 45'	-4.9	30'	0%	7° W	NM		Venus
17th	Cap	20h 15m	-18° 36'	-4.7	30'	0%	5° E	NM		Venus
18th	Cap	21h 04m	-16° 36'	-5.6	30'	2%	15° E	NM		
19th	Cap	21h 53m	-13° 51'	-6.4	30'	5%	26° E	NM		
20th	Aqr	22h 41m	-10° 30'	-7.2	30'	10%	38° E	NM		Neptune

Mercury and Venus

Mercury	Mercury	Venus
13th	17th	15th

Mercury

Date	Con	R.A.	Dec	Mag	Diam	Ill.	Elong	AM/PM		Close To
11th	Sgr	18h 02m	-22° 57'	-0.3	6''	80%	21° W	AM		Saturn
13th	Sgr	18h 14m	-23° 12'	-0.3	6''	83%	20° W	AM		Saturn
15th	Sgr	18h 26m	-23° 21'	-0.3	5''	85%	19° W	AM		Moon, Saturn
17th	Sgr	18h 38m	-23° 27'	-0.3	5''	86%	19° W	AM		Saturn
19th	Sgr	18h 51m	-23° 27'	-0.3	5''	88%	18° W	AM		Saturn

Venus

Date	Con	R.A.	Dec	Mag	Diam	Ill	Elong	AM/PM		Close To
11th	Sgr	19h 35m	-22° 30'	-3.9	10''	100%	1° E	NV		
13th	Sgr	19h 45m	-22° 9'	-3.9	10''	100%	1° E	NV		
15th	Sgr	19h 56m	-21° 45'	-3.9	10''	100%	2° E	NV		
17th	Sgr	20h 07m	-21° 18'	-3.9	10''	100%	2° E	NV		Moon
19th	Cap	20h 17m	-20° 48'	-3.9	10''	100%	3° E	NV		

Mars and the Outer Planets

Mars
15th

Jupiter
15th

Saturn
15th

Mars

Date	Con	R.A.	Dec	Mag	Diam	Ill	Elong	AM/PM		Close To
10th	Lib	15h 12m	-16° 57'	1.4	5''	93%	60° W	AM		Moon, Jupiter
15th	Lib	15h 24m	-17° 48'	1.4	5''	92%	62° W	AM		Jupiter
20th	Lib	15h 37m	-18° 36'	1.3	5''	92%	64° W	AM		Jupiter

The Outer Planets

Planet	Date	Con	R.A.	Dec	Mag	Diam	Elong	AM/PM		Close To
Jupiter	15th	Lib	15h 08m	-16° 30'	-1.9	34''	66° W	AM		Mars
Saturn	15th	Sgr	18h 13m	-22° 30'	0.5	15''	22° W	AM		Moon, Mercury
Uranus	15th	Psc	01h 32m	-9° 0'	5.8	4''	89° E	PM		
Neptune	15th	Aqr	22h 56m	-7° 48'	7.9	2''	47° E	PM		

Events

Date	Time (UT)	Event
11th	04:36	The waning crescent Moon appears north of Jupiter. (Morning sky.)
	11:47	The waning crescent Moon appears north of Mars. (Morning sky.)
13th	N/A	Good opportunity to see Earthshine on the waning crescent Moon. (Morning sky.)
	06:47	Mercury appears 0.6° south of Saturn. (Morning sky.)
15th	00:55	The waning crescent Moon appears north of Saturn. (Morning sky.)
	05:21	The waning crescent Moon appears north of Mercury. (Morning sky.)
17th	02:18	New Moon. (Not visible.)
20th	N/A	Good opportunity to see Earthshine on the waxing crescent Moon. (Evening sky.)
	21:26	The waxing crescent Moon appears south of Neptune. (Evening sky.)
	06:21	Venus is at superior conjunction with the Sun. (Not visible.)
	06:21	Venus is at superior conjunction with the Sun. (Not visible.)

Planet Locations – January 15th

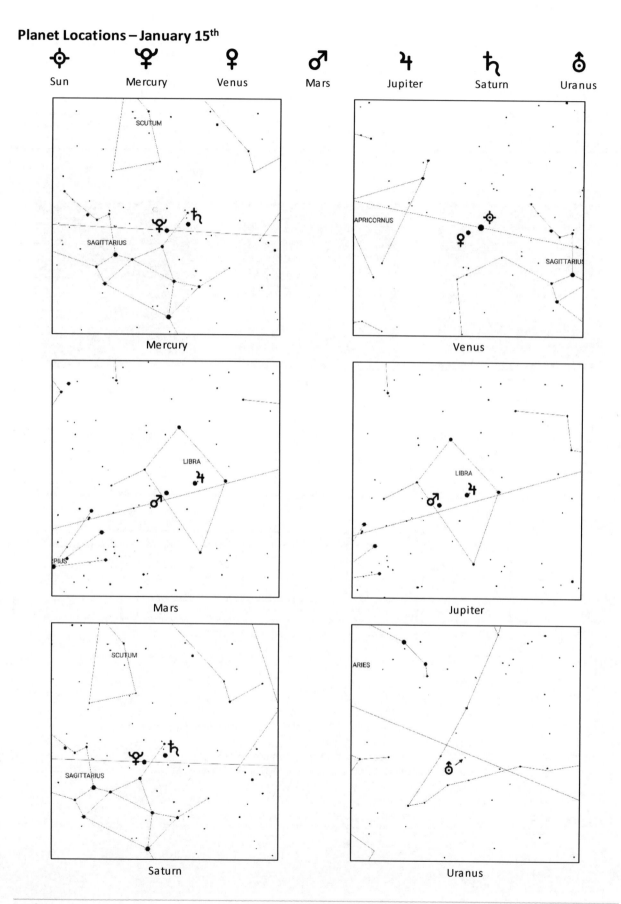

Sun	Mercury	Venus	Mars	Jupiter	Saturn	Uranus

The Moon

| 21st | 23rd | 25th | 27th | 29th | 31st |

Date	Con	R.A.	Dec	Mag	Diam	Ill.	Elong	Phase	Close To
21st	Aqr	23h 29m	-6° 39'	-7.9	31'	17%	49° E	+Cr	
22nd	Aqr	00h 17m	-2° 27'	-8.5	31'	26%	61° E	+Cr	
23rd	Cet	01h 05m	1° 57'	-9.1	31'	35%	73° E	+Cr	
24th	Psc	01h 55m	6° 21'	-9.6	31'	46%	85° E	FQ	Uranus
25th	Ari	02h 47m	10° 36'	-10.1	32'	56%	97° E	FQ	
26th	Tau	03h 42m	14° 21'	-10.5	32'	67%	110° E	+G	Pleiades, Hyades
27th	Tau	04h 40m	17° 21'	-11.0	32'	78%	123° E	+G	Hyades, Aldebaran
28th	Tau	05h 42m	19° 21'	-11.4	33'	87%	137° E	+G	
29th	Gem	06h 45m	20° 3'	-11.8	33'	94%	151° E	+G	
30th	Gem	07h 49m	19° 18'	-12.2	33'	98%	165° E	FM	
31st	Cnc	08h 52m	17° 8'	-12.7	33'	100%	179° E	FM	Praesepe

Mercury and Venus

Mercury
23rd

Mercury
27th

Venus
25th

Mercury

Date	Con	R.A.	Dec	Mag	Diam	Ill.	Elong	AM/PM	Close To
21st	Sgr	19h 04m	-23° 24'	-0.3	5''	89%	17° W	AM	
23rd	Sgr	19h 17m	-23° 15'	-0.4	5''	91%	16° W	AM	
25th	Sgr	19h 30m	-23° 3'	-0.4	5''	92%	15° W	NV	
27th	Sgr	19h 43m	-22° 42'	-0.5	5''	93%	14° W	NV	
29th	Sgr	19h 56m	-22° 18'	-0.5	5''	94%	13° W	NV	
31st	Cap	20h 10m	-21° 48'	-0.6	5''	95%	12° W	NV	

Venus

Date	Con	R.A.	Dec	Mag	Diam	Ill	Elong	AM/PM	Close To
21st	Cap	20h 28m	-20° 15'	-3.9	10''	100%	3° E	NV	
23rd	Cap	20h 38m	-19° 42'	-3.9	10''	100%	4° E	NV	
25th	Cap	20h 48m	-19° 6'	-3.9	10''	100%	4° E	NV	
27th	Cap	20h 59m	-18° 27'	-3.9	10''	100%	5° E	NV	
29th	Cap	21h 09m	-17° 45'	-3.9	10''	100%	5° E	NV	
31st	Cap	21h 19m	-17° 3'	-3.9	10''	100%	6° E	NV	

Mars and the Outer Planets

Mars
25th

Jupiter
25th

Saturn
25th

Mars

Date	Con	R.A.	Dec	Mag	Diam	Ill	Elong	AM/PM	Close To
20th	Lib	15h 37m	-18° 36'	1.3	5''	92%	64° W	AM	Jupiter
25th	Lib	15h 49m	-19° 18'	1.3	5''	91%	66° W	AM	Jupiter
30th	Sco	16h 02m	-20° 0'	1.2	6''	91%	68° W	AM	Antares

The Outer Planets

Planet	Date	Con	R.A.	Dec	Mag	Diam	Elong	AM/PM	Close To
Jupiter	25th	Lib	15h 14m	-16° 48'	-1.9	35''	75° W	AM	Mars
Saturn	25th	Sgr	18h 18m	-22° 30'	0.6	15''	31° W	AM	
Uranus	25th	Psc	01h 33m	-9° 6'	5.8	4''	79° E	PM	
Neptune	25th	Aqr	22h 57m	-7° 39'	7.9	2''	37° E	PM	

Events

Date	Time (UT)	Event
24th	02:29	The almost first quarter Moon appears south of Uranus. (Evening sky.)
	22:21	First Quarter Moon. (Evening sky.)
27th	09:33	The waxing gibbous Moon appears north of the bright star Aldebaran. (Taurus, evening sky.)
31st	13:27	Full Moon. (All night.)
	13:27	Total lunar eclipse. (North-east Africa, Asia, Australia, Eastern Europe, Indian Ocean, North America and the Pacific Ocean.)

Planet Locations – January 25th

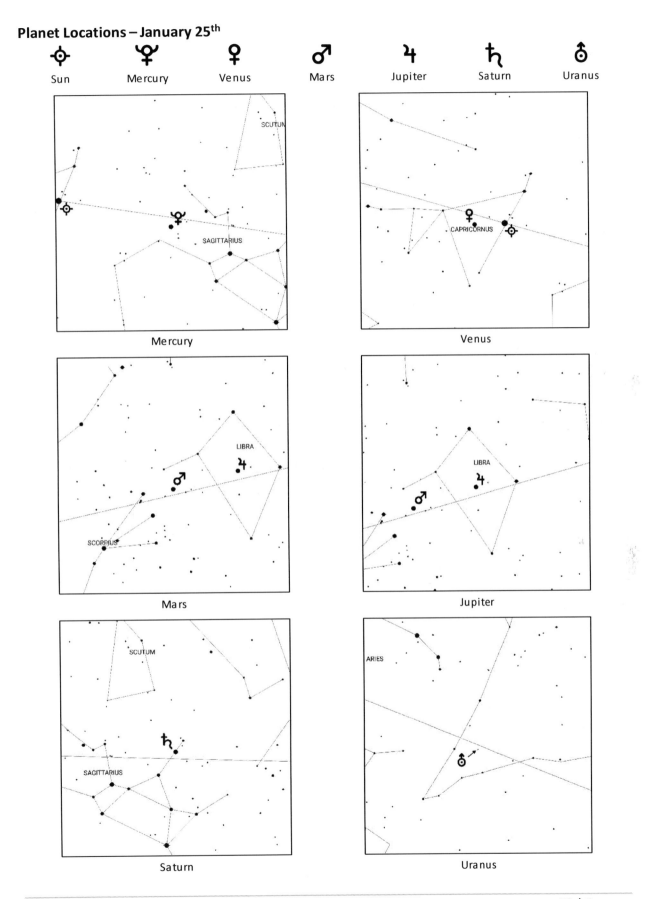

February 1st to 10th

The Moon

| | 1st | | 3rd | | 5th | | 7th | | 9th |

Date	Con	R.A.	Dec	Mag	Diam	Ill.	Elong	Phase	Close To
1st	Leo	09h 53m	13° 54'	-12.3	32'	99%	167° W	FM	Regulus
2nd	Leo	10h 50m	9° 51'	-11.9	32'	95%	153° W	-G	Regulus
3rd	Vir	11h 45m	5° 21'	-11.5	32'	88%	140° W	-G	
4th	Vir	12h 36m	0° 42'	-11.1	31'	81%	127° W	-G	Spica
5th	Vir	13h 26m	-3° 48'	-10.7	31'	71%	115° W	-G	Spica
6th	Vir	14h 15m	-8° 3'	-10.3	30'	62%	103° W	LQ	Spica
7th	Lib	15h 04m	-11° 45'	-9.9	30'	52%	92° W	LQ	Jupiter
8th	Lib	15h 52m	-14° 54'	-9.4	30'	42%	81° W	LQ	Mars, Jupiter, Antares
9th	Oph	16h 41m	-17° 24'	-9.0	30'	33%	70° W	-Cr	Mars, Antares
10th	Oph	17h 30m	-19° 3'	-8.4	30'	24%	59° W	-Cr	

Mercury and Venus

Mercury
3rd

Mercury
7th

Venus
5th

Mercury

Date	Con	R.A.	Dec	Mag	Diam	Ill.	Elong	AM/PM	Close To
1st	Cap	20h 17m	-21° 30'	-0.6	5''	96%	11° W	NV	
3rd	Cap	20h 30m	-20° 54'	-0.7	5''	96%	10° W	NV	
5th	Cap	20h 44m	-20° 9'	-0.8	5''	97%	9° W	NV	
7th	Cap	20h 58m	-19° 18'	-0.9	5''	98%	7° W	NV	
9th	Cap	21h 11m	-18° 24'	-1.0	5''	98%	6° W	NV	

Venus

Date	Con	R.A.	Dec	Mag	Diam	Ill	Elong	AM/PM	Close To
1st	Cap	21h 24m	-16° 39'	-3.9	10''	100%	6° E	NV	
3rd	Cap	21h 34m	-15° 57'	-3.9	10''	100%	6° E	NV	
5th	Cap	21h 44m	-15° 6'	-3.9	10''	99%	7° E	NV	
7th	Cap	21h 54m	-14° 18'	-3.9	10''	99%	7° E	NV	
9th	Aqr	22h 03m	-13° 27'	-3.9	10''	99%	8° E	NV	

Mars and the Outer Planets

Mars
5th

Jupiter
5th

Saturn
5th

Mars

Date	Con	R.A.	Dec	Mag	Diam	Ill	Elong	AM/PM	Close To
1st	Sco	16h 07m	-20° 15'	1.2	6''	91%	69° W	AM	Antares
5th	Sco	16h 17m	-20° 45'	1.1	6''	91%	71° W	AM	Antares
10th	Oph	16h 30m	-21° 18'	1.1	6''	90%	73° W	AM	Antares

The Outer Planets

Planet	Date	Con	R.A.	Dec	Mag	Diam	Elong	AM/PM	Close To
Jupiter	5th	Lib	15h 18m	-17° 6'	-2.0	36''	85° W	AM	
Saturn	5th	Sgr	18h 23m	-22° 27'	0.6	15''	41° W	AM	
Uranus	5th	Psc	01h 34m	9° 12'	5.8	4''	68° E	PM	
Neptune	5th	Aqr	22h 59m	-7° 33'	8.0	2''	26° E	PM	

Events

Date	Time (UT)	Event
1st	17:00	The just-past full Moon appears north of the bright star Regulus. (Leo, all night.)
3rd	16:43	Dwarf planet Ceres is at opposition. (All night.)
5th	12:51	The waning gibbous Moon appears north of the bright star Spica. (Virgo, morning sky.)
7th	15:55	Last Quarter Moon. (Morning sky.)
	19:15	The last quarter Moon appears to the north of Jupiter. (Morning sky.)
9th	03:30	The waning crescent Moon appears to the north of Mars. (Morning sky.)
	05:43	The waning crescent Moon appears to the north of the bright star Antares. (Scorpius, morning sky.)
10th	04:38	Mars appears 5.2° north of the bright star Antares. (Scorpius, morning sky.)

Planet Locations – February 5th

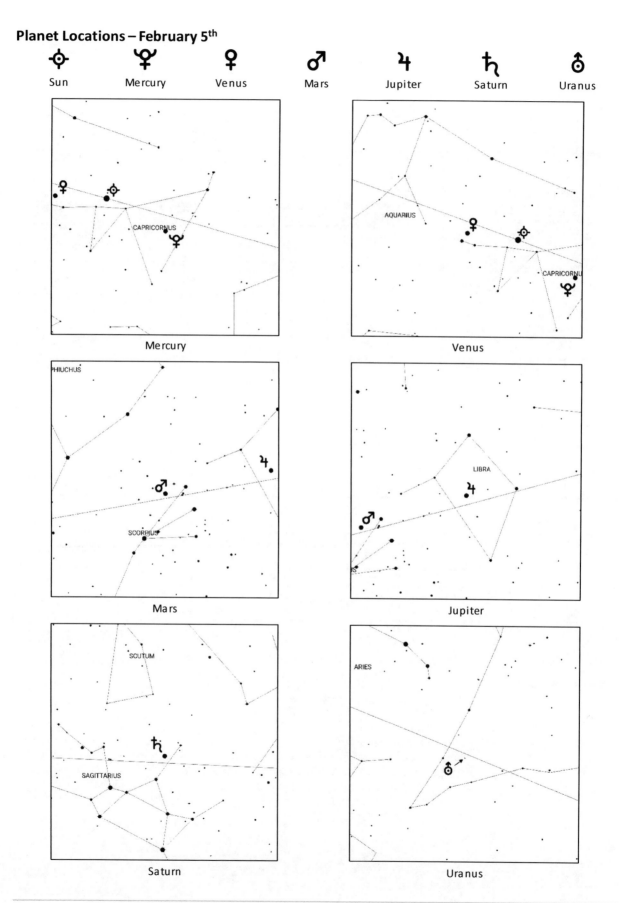

The Moon

| 11th | 13th | 15th | 17th | 19th |

Date	Con	R.A.	Dec	Mag	Diam	Ill.	Elong	Phase	Close To
11th	Sgr	18h 20m	-19° 54'	-7.8	30'	17%	48° W	-Cr	Saturn
12th	Sgr	19h 10m	-19° 54'	-7.2	30'	10%	37° W	NM	Saturn
13th	Sgr	20h 01m	-19° 0'	-6.4	30'	5%	26° W	NM	
14th	Cap	20h 51m	-17° 12'	-5.6	30'	2%	15° W	NM	
15th	Cap	21h 40m	-14° 39'	-4.7	30'	0%	4° W	NM	Mercury
16th	Aqr	22h 29m	-11° 24'	-5.0	31'	0%	7° E	NM	Mercury, Venus
17th	Aqr	23h 18m	-7° 36'	-5.9	31'	3%	19° E	NM	Venus, Neptune
18th	Psc	00h 06m	-3° 27'	-6.7	31'	7%	30° E	NM	
19th	Cet	00h 54m	0° 57'	-7.5	31'	13%	42° E	+Cr	
20th	Psc	01h 42m	5° 24'	-8.2	32'	21%	54° E	+Cr	Uranus

Mercury and Venus

Mercury
13th

Mercury
17th

Venus
15th

Mercury

Date	Con	R.A.	Dec	Mag	Diam	Ill.	Elong	AM/PM	Close To
11th	Cap	21h 25m	-17° 21'	-1.1	5''	99%	5° W	NV	
13th	Cap	21h 39m	-16° 15'	-1.3	5''	99%	4° W	NV	
15th	Cap	21h 53m	-15° 3'	-1.4	5''	100%	3° W	NV	Moon, Venus
17th	Aqr	22h 06m	-13° 45'	-1.5	5''	100%	2° W	NV	Venus
19th	Aqr	22h 20m	-12° 21'	-1.6	5''	100%	3° E	NV	Venus

Venus

Date	Con	R.A.	Dec	Mag	Diam	Ill	Elong	AM/PM	Close To
11th	Aqr	22h 13m	-12° 36'	-3.9	10''	99%	8° E	NV	
13th	Aqr	22h 23m	-11° 42'	-3.9	10''	99%	9° E	NV	
15th	Aqr	22h 32m	-10° 48'	-3.9	10''	99%	9° E	NV	Mercury
17th	Aqr	22h 42m	-9° 51'	-3.9	10''	99%	10° E	NV	Moon, Mercury, Neptune
19th	Aqr	22h 51m	-8° 54'	-3.9	10''	99%	10° E	NV	Mercury, Neptune

Mars and the Outer Planets

Mars
15th

Jupiter
15th

Saturn
15th

Mars

Date	Con	R.A.	Dec	Mag	Diam	Ill	Elong	AM/PM	Close To
10th	Oph	16h 30m	-21° 18'	1.1	6''	90%	73° W	AM	Antares
15th	Oph	16h 43m	-21° 48'	1.0	6''	90%	75° W	AM	Antares
20th	Oph	16h 56m	-22° 12'	0.9	6''	90%	77° W	AM	Antares

The Outer Planets

Planet	Date	Con	R.A.	Dec	Mag	Diam	Elong	AM/PM	Close To
Jupiter	15th	Lib	15h 20m	-17° 18'	-2.1	37''	94° W	AM	
Saturn	15th	Sgr	18h 27m	-22° 24'	0.6	16''	51° W	AM	
Uranus	15th	Psc	01h 35m	9° 18'	5.9	3''	59° E	PM	
Neptune	15th	Aqr	23h 00m	-7° 24'	8.0	2''	17° E	NV	

Events

Date	Time (UT)	Event
11th	16:14	The waning crescent Moon appears to the north of Saturn. (Morning sky.)
12th	N/A	Good opportunity to see Earthshine on the waning crescent Moon. (Morning sky.)
15th	20:52	Partial solar eclipse. (Atlantic Ocean, Pacific Ocean and South America.)
	21:06	New Moon
17th	12:11	Mercury is at superior conjunction with the Sun. (Not visible.)
18th	N/A	Good opportunity to see Earthshine on the waxing crescent Moon. (Evening sky.)
20th	06:45	The waxing crescent Moon appears to the south of Uranus. (Evening sky.)

Planet Locations – February 14th

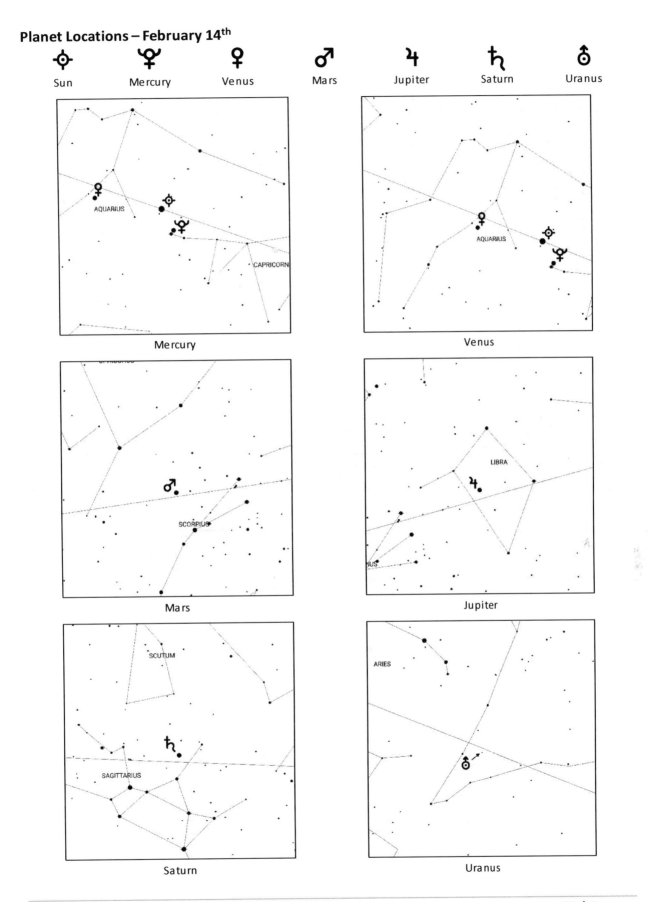

February 21st – 28th

The Moon

21st   23rd   25th   27th

Date	Con	R.A.	Dec	Mag	Diam	Ill.	Elong	Phase		Close To
21st	Cet	02h 35m	9° 39'	-8.8	32'	30%	67° E	+Cr		
22nd	Tau	03h 28m	13° 27'	-9.4	32'	41%	79° E	FQ		Pleiades
23rd	Tau	04h 24m	16° 38'	-9.9	32'	52%	92° E	FQ		Hyades, Aldebaran
24th	Tau	05h 22m	18° 53'	-10.4	32'	63%	105° E	FQ		Aldebaran
25th	Gem	06h 23m	19° 57'	-10.8	32'	74%	118° E	+G		
26th	Gem	07h 24m	19° 45'	-11.2	32'	84%	132° E	+G		
27th	Cnc	08h 26m	18° 12'	-11.6	32'	91%	146° E	+G		Praesepe
28th	Leo	09h 27m	15° 27'	-12.0	32'	97%	159° E	FM		Praesepe, Regulus

Mercury and Venus

Mercury   Mercury   Venus

23rd   27th   24th

Mercury

Date	Con	R.A.	Dec	Mag	Diam	Ill.	Elong	AM/PM		Close To
21st	Aqr	22h 34m	-10° 51'	-1.5	5''	99%	4° E	NV		Venus
23rd	Aqr	22h 48m	-9° 18'	-1.5	5''	98%	5° E	NV		Venus, Neptune
25th	Aqr	23h 02m	-7° 39'	-1.4	5''	97%	7° E	NV		Venus, Neptune
27th	Aqr	23h 15m	-5° 57'	-1.4	5''	95%	9° E	NV		Venus, Neptune

Venus

Date	Con	R.A.	Dec	Mag	Diam	Ill	Elong	AM/PM	Close To
21st	Aqr	23h 00m	-7° 57'	-3.9	10''	98%	11° E	NV	Mercury, Neptune
23rd	Aqr	23h 10m	-6° 57'	-3.9	10''	98%	11° E	NV	Mercury, Neptune
25th	Aqr	23h 19m	-5° 57'	-3.9	10''	98%	11° E	NV	Mercury, Neptune
27th	Aqr	23h 28m	-4° 57'	-3.9	10''	98%	12° E	NV	Mercury

Mars and the Outer Planets

Mars
24th

Jupiter
24th

Saturn
24th

Mars

Date	Con	R.A.	Dec	Mag	Diam	Ill	Elong	AM/PM	Close To
20th	Oph	16h 56m	-22° 12'	0.9	6''	90%	77° W	AM	Antares
25th	Oph	17h 09m	-22° 33'	0.9	7''	89%	79° W	AM	

The Outer Planets

Planet	Date	Con	R.A.	Dec	Mag	Diam	Elong	AM/PM	Close To
Jupiter	25th	Lib	15h 22m	17° 23'	-2.1	39''	104° W	AM	
Saturn	25th	Sgr	18h 31m	-22° 21'	0.6	16''	60° W	AM	
Uranus	25th	Psc	01h 36m	9° 27'	5.9	3''	49° E	PM	
Neptune	25th	Aqr	23h 01m	-7° 15'	8.0	2''	7° E	NV	Mercury, Venus

Events

Date	Time (UT)	Event
22nd	22:00	The nearly first quarter Moon appears to the south of the Pleiades star cluster. (Taurus, evening sky.)
23rd	08:10	First Quarter Moon. (Evening sky.)
	16:19	The first quarter Moon appears to the north of the bright star Aldebaran. (Taurus, evening sky.)

Planet Locations – February 24th

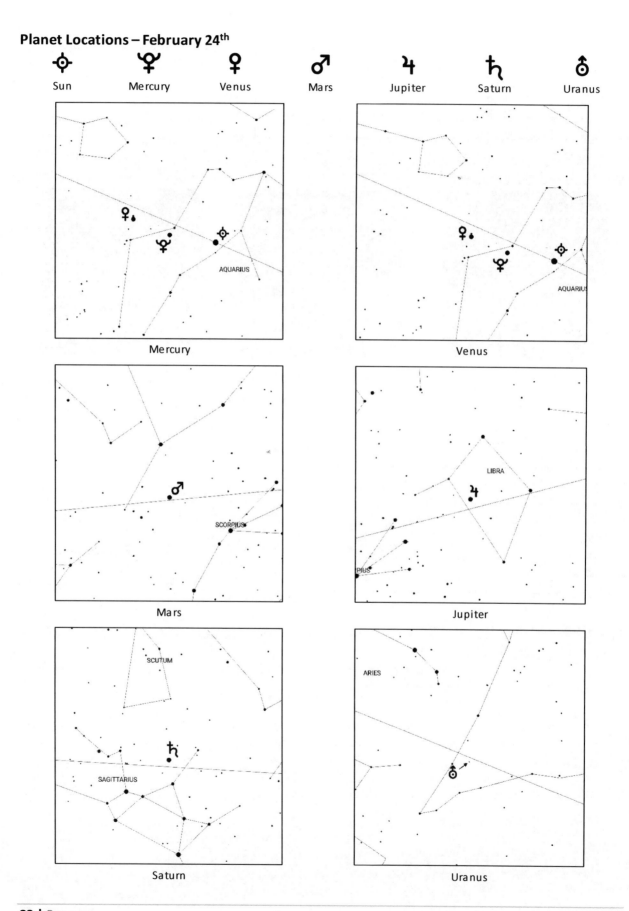

The Moon

1st 3rd 5th 7th 9th

Date	Con	R.A.	Dec	Mag	Diam	Ill.	Elong	Phase	Close To
1st	Leo	10h 25m	11° 45'	-12.5	32'	100%	173° E	FM	Regulus
2nd	Leo	11h 20m	7° 21'	-12.5	32'	100%	173° W	FM	
3rd	Vir	12h 14m	2° 42'	-12.1	31'	97%	160° W	FM	
4th	Vir	13h 06m	-2° 0'	-11.7	31'	92%	148° W	-G	Spica
5th	Vir	13h 56m	-6° 30'	-11.3	31'	86%	135° W	-G	Spica
6th	Lib	14h 46m	-10° 30'	-11.0	30'	78%	124° W	-G	Jupiter
7th	Lib	15h 35m	-14° 0'	-10.6	30'	69%	112° W	-G	Jupiter
8th	Sco	16h 24m	-16° 45'	-10.2	30'	59%	101° W	LQ	Antares
9th	Oph	17h 14m	-18° 42'	-9.8	30'	50%	90° W	LQ	Mars, Antares
10th	Sgr	18h 04m	-19° 51'	-9.4	30'	41%	79° W	LQ	Mars, Saturn

Mercury and Venus

Mercury 3rd Mercury 7th Venus 5th

Mercury

Date	Con	R.A.	Dec	Mag	Diam	Ill.	Elong	AM/PM	Close To
1st	Aqr	23h 29m	-4° 12'	-1.3	5''	92%	10° E	NV	Venus
3rd	Psc	23h 42m	-2° 24'	-1.2	6''	88%	12° E	NV	Venus
5th	Psc	23h 55m	0° 36'	-1.2	6''	83%	14° E	NV	Venus
7th	Psc	00h 07m	1° 8'	-1.1	6''	77%	15° E	NV	Venus
9th	Psc	00h 18m	2° 51'	-0.9	6''	70%	17° E	PM	Venus

Venus

Date	Con	R.A.	Dec	Mag	Diam	Ill	Elong	AM/PM	Close To
1st	Aqr	23h 37m	-3° 57'	-3.9	10''	98%	12° E	NV	Mercury
3rd	Psc	23h 46m	-2° 57'	-3.9	10''	98%	13° E	NV	Mercury
5th	Psc	23h 55m	-1° 54'	-3.9	10''	97%	13° E	NV	Mercury
7th	Psc	00h 04m	-1° 0'	-3.9	10''	97%	14° E	NV	Mercury
9th	Psc	00h 13m	0° 9'	-3.9	10''	97%	14° E	NV	Mercury

Mars and the Outer Planets

Mars
5th

Jupiter
5th

Saturn
5th

Mars

Date	Con	R.A.	Dec	Mag	Diam	Ill	Elong	AM/PM	Close To
1st	Oph	17h 19m	-22° 51'	0.8	7''	89%	80° W	AM	
5th	Oph	17h 29m	-23° 3'	0.7	7''	89%	82° W	AM	
10th	Oph	17h 42m	-23° 15'	0.7	7''	89%	84° W	AM	Moon

The Outer Planets

Planet	Date	Con	R.A.	Dec	Mag	Diam	Elong	AM/PM	Close To
Jupiter	5th	Lib	15h 25m	-17° 24'	-2.2	40''	112° W	AM	
Saturn	5th	Sgr	18h 33m	-22° 21'	0.6	16''	67° W	AM	
Uranus	5th	Psc	01h 38m	9° 36'	5.9	3''	41° E	PM	
Neptune	5th	Aqr	23h 02m	-7° 9'	8.0	2''	1° W	NV	

Events

Date	Time (UT)	Event
1st	06:43	The nearly full Moon appears north of the bright star Regulus. (All night.)
2nd	00:52	Full Moon. (All night.)
4th	19:29	The waning gibbous Moon appears north of the bright star Spica. (Virgo, morning sky.)
	23:23	Neptune is in conjunction with the Sun. (Not visible.)
7th	08:34	The waning gibbous Moon appears to the north of Jupiter. (Morning sky.)
8th	15:18	The nearly last quarter Moon appears to the north of the bright star Antares. (Scorpius, morning sky.)
10th	09:05	Jupiter is stationary prior to beginning retrograde motion. (Morning sky.)
	11:21	Last Quarter Moon. (Morning sky.)
	22:51	The last quarter Moon appears to the north of Mars. (Morning sky.)

Planet Locations – March 5th

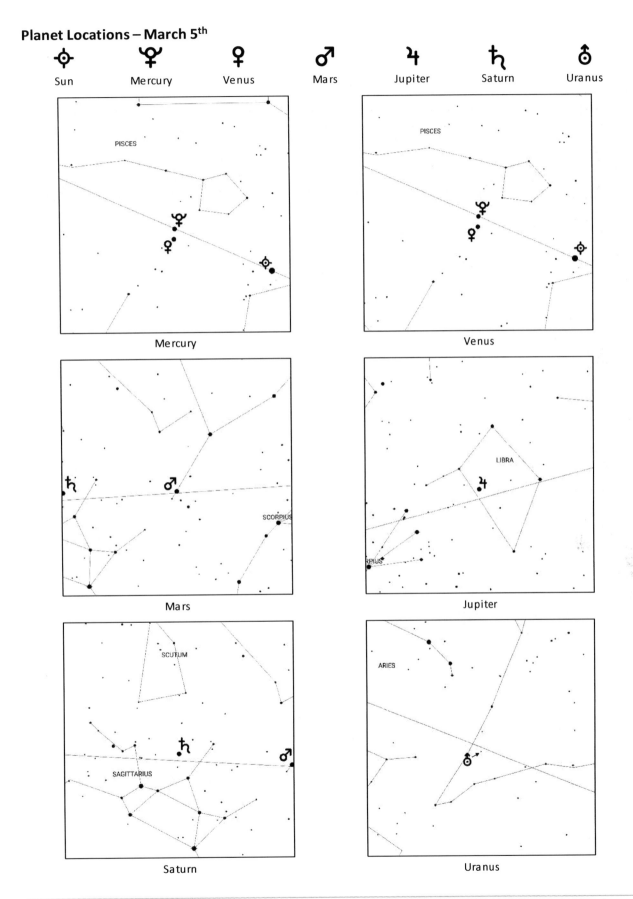

March 11th – 20th

The Moon

| | 11th | | 13th | | 15th | | 17th | | 19th |

Date	Con	R.A.	Dec	Mag	Diam	Ill.	Elong	Phase	Close To
11th	Sgr	18h 55m	-20° 6'	-8.9	30'	31%	68° W	-Cr	Saturn
12th	Sgr	19h 45m	-19° 27'	-8.3	30'	23%	57° W	-Cr	
13th	Cap	20h 35m	-17° 54'	-7.7	30'	16%	46° W	-Cr	
14th	Cap	21h 25m	-15° 36'	-7.1	30'	9%	35° W	NM	
15th	Aqr	22h 14m	-12° 30'	-6.3	31'	4%	24° W	NM	
16th	Aqr	23h 03m	-8° 48'	-5.4	31'	1%	13° W	NM	Neptune
17th	Aqr	23h 51m	-4° 39'	-4.6	31'	0%	4° W	NM	
18th	Cet	00h 41m	0° 12'	-5.4	32'	1%	12° E	NM	Mercury, Venus
19th	Psc	01h 31m	4° 21'	-6.3	32'	4%	24° E	NM	Mercury, Venus, Uranus
20th	Cet	02h 22m	8° 45'	-7.1	32'	10%	37° E	NM	

Mercury and Venus

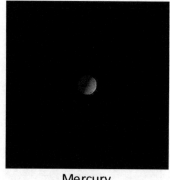  

Mercury
13th

Mercury
17th

Venus
15th

Mercury

Date	Con	R.A.	Dec	Mag	Diam	Ill.	Elong	AM/PM	Close To
11th	Psc	00h 29m	4° 15'	-0.7	7''	62%	18° E	PM	Venus
13th	Psc	00h 38m	5° 51'	-0.5	7''	54%	18° E	PM	Venus
15th	Psc	00h 45m	7° 5'	-0.2	7''	45%	18° E	PM	Venus
17th	Psc	00h 51m	8° 9'	0.1	8''	37%	18° E	PM	Venus
19th	Psc	00h 55m	8° 57'	0.5	8''	29%	18° E	PM	Moon, Venus

Venus

Date	Con	R.A.	Dec	Mag	Diam	Ill	Elong	AM/PM	Close To
11th	Psc	00h 22m	1° 8'	-3.9	10''	97%	15° E	NV	Mercury
13th	Cet	00h 32m	2° 12'	-3.9	10''	97%	15° E	NV	Mercury
15th	Psc	00h 41m	3° 12'	-3.9	10''	96%	16° E	PM	Mercury
17th	Psc	00h 50m	4° 15'	-3.9	10''	96%	16° E	PM	Mercury
19th	Psc	00h 59m	5° 15'	-3.9	10''	96%	17° E	PM	Moon, Mercury

Mars and the Outer Planets

Mars
15th

Jupiter
15th

Saturn
15th

Mars

Date	Con	R.A.	Dec	Mag	Diam	Ill	Elong	AM/PM	Close To
10th	Oph	17h 42m	-23° 15'	0.7	7''	89%	84° W	AM	Moon
15th	Sgr	17h 55m	-23° 24'	0.6	7''	89%	86° W	AM	
20th	Sgr	18h 07m	-23° 30'	0.5	8''	88%	88° W	AM	Saturn

The Outer Planets

Planet	Date	Con	R.A.	Dec	Mag	Diam	Elong	AM/PM	Close To
Jupiter	15th	Lib	15h 24m	-17° 21'	-2.3	41''	122° W	AM	
Saturn	15th	Sgr	18h 36m	-22° 18'	0.6	16''	77° W	AM	
Uranus	15th	Psc	01h 40m	9° 45'	5.9	3''	32° E	PM	
Neptune	15th	Aqr	23h 04m	-7° 0'	8.0	2''	11° W	NV	Moon

Events

Date	Time (UT)	Event
11th	00:32	The waning crescent Moon appears to the north of Saturn. (Morning sky.)
12th	02:26	The waning crescent Moon appears to the north of the dwarf planet Pluto. (Morning sky.)
13th	N/A	Good opportunity to see earthshine on the waning crescent Moon. (Morning sky.)
15th	14:58	Mercury is at greatest eastern elongation from the Sun. (Evening sky.)
17th	13:12	New Moon. (Not visible.)
18th	00:54	Mercury appears 3.9° north of Venus. (Evening sky.)
	20:02	The just-past new Moon appears to the south of Mercury. (Evening sky.)
	20:56	The just-past new Moon appears to the south of Venus. (Evening sky.)
19th	18:10	The waxing crescent Moon appears to the south of Uranus. (Evening sky.)
20th	16:16	Vernal equinox. Spring begins in the northern hemisphere, autumn begins in the south.
	N/A	Good opportunity to see Earthshine on the waxing crescent Moon.

Planet Locations – March 15th

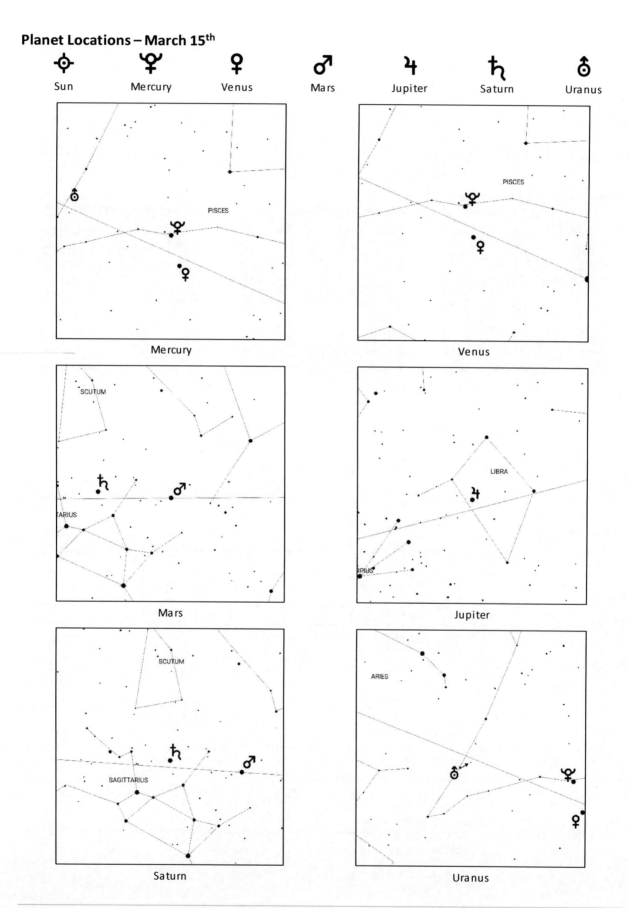

The Moon

| 21st | 23rd | 25th | 27th | 29th | 31st |

Date	Con	R.A.	Dec	Mag	Diam	Ill.	Elong	Phase	Close To
21st	Ari	03h 15m	12° 48'	-7.9	32'	17%	49° E	+Cr	Pleiades
22nd	Tau	04h 11m	16° 8'	-8.6	32'	27%	62° E	+Cr	Pleiades, Hyades
23rd	Tau	05h 08m	18° 36'	-9.2	32'	37%	75° E	+Cr	Hyades, Aldebaran
24th	Ori	06h 08m	20° 0'	-9.7	32'	48%	88° E	FQ	
25th	Gem	07h 08m	20° 6'	-10.2	32'	60%	101° E	FQ	
26th	Cnc	08h 08m	18° 57'	-10.7	32'	71%	114° E	+G	Praesepe
27th	Cnc	09h 07m	16° 33'	-11.1	32'	81%	128° E	+G	Praesepe
28th	Leo	10h 05m	13° 12'	-11.5	32'	89%	141° E	+G	Regulus
29th	Leo	11h 00m	9° 5'	-11.9	32'	95%	154° E	FM	
30th	Vir	11h 53m	4° 33'	-12.3	31'	99%	166° E	FM	
31st	Vir	12h 45m	0° 12'	-12.6	31'	100%	176° E	FM	Spica

Mercury and Venus

Mercury 23rd   Mercury 27th   Venus 25th

Mercury

Date	Con	R.A.	Dec	Mag	Diam	Ill.	Elong	AM/PM	Close To
21st	Psc	00h 57m	9° 27'	1.1	9''	21%	16° E	PM	Venus
23rd	Psc	00h 57m	9° 42'	1.7	9''	15%	15° E	NV	Venus
25th	Psc	00h 56m	9° 39'	2.4	10''	9%	12° E	NV	Venus
27th	Psc	00h 52m	9° 18'	3.2	10''	5%	9° E	NV	
29th	Psc	00h 48m	8° 42'	4.1	11''	2%	7° E	NV	
31st	Psc	00h 43m	7° 54'	4.9	11''	1%	4° E	NV	

Venus

Date	Con	R.A.	Dec	Mag	Diam	Ill	Elong	AM/PM	Close To
21st	Psc	01h 08m	6° 15'	-3.9	10''	96%	17° E	PM	Mercury
23rd	Psc	01h 17m	7° 15'	-3.9	10''	95%	18° E	PM	Mercury
25th	Psc	01h 26m	8° 15'	-3.9	10''	95%	18° E	PM	Mercury, Uranus
27th	Psc	01h 35m	9° 12'	-3.9	11''	95%	19° E	PM	Uranus
29th	Psc	01h 45m	10° 12'	-3.9	11''	95%	19° E	PM	Uranus
31st	Ari	01h 54m	11° 9'	-3.9	11''	94%	20° E	PM	Uranus

Mars and the Outer Planets

Mars
25th

Jupiter
25th

Saturn
25th

Mars

Date	Con	R.A.	Dec	Mag	Diam	Ill	Elong	AM/PM	Close To
20th	Sgr	18h 07m	-23° 30'	0.5	8''	88%	88° W	AM	Saturn
25th	Sgr	18h 19m	-23° 33'	0.4	8''	88%	90° W	AM	Saturn
30th	Sgr	18h 32m	-23° 33'	0.3	8''	88%	93° W	AM	Saturn

The Outer Planets

Planet	Date	Con	R.A.	Dec	Mag	Diam	Elong	AM/PM	Close To
Jupiter	25th	Lib	15h 23m	-17° 15'	-2.3	42''	132° W	AM	
Saturn	25th	Sgr	18h 38m	-22° 15'	0.5	16''	86° W	AM	Mars
Uranus	25th	Psc	01h 42m	9° 57'	5.9	3''	22° E	PM	Venus
Neptune	25th	Aqr	23h 05m	-6° 51'	8.0	2''	20° W	AM	

Events

Date	Time (UT)	Event
22nd	02:24	The waxing crescent Moon appears south of the Pleaides star cluster. (Taurus, evening sky.)
	17:11	Mercury is stationary prior to beginning retrograde motion. (Not visible.)
23rd	22:08	The nearly first quarter Moon appears north of the bright star Aldebaran. (Taurus, evening sky.)
24th	15:36	First Quarter Moon. (Evening sky.)
28th	18:51	The waxing gibbous Moon appears north of the bright star Regulus. (Leo, evening sky.)
29th	00:11	Venus appears 0.1° south of Uranus. (Evening sky.)
31st	12:38	Full Moon. (All night.)

Planet Locations – March 25th

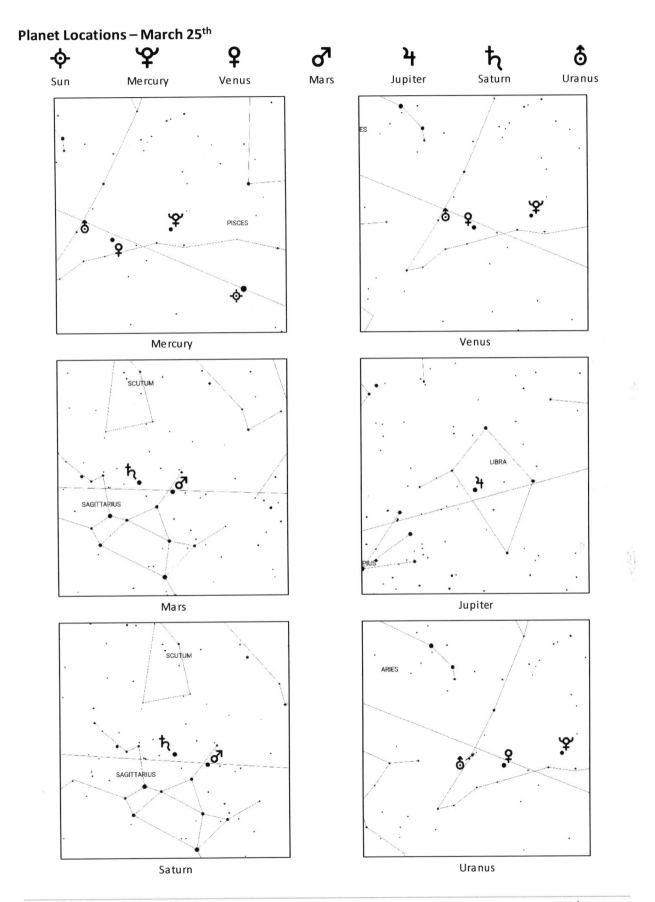

April 1st to 10th

The Moon

| 1st | 3rd | 5th | 7th | 9th |

Date	Con	R.A.	Dec	Mag	Diam	Ill.	Elong	Phase	Close To
1st	Vir	13h 36m	-4° 48'	-12.3	31'	99%	0° W	FM	Spica
2nd	Lib	14h 26m	-9° 6'	-11.9	30'	95%	0° W	-G	
3rd	Lib	15h 16m	-12° 54'	-11.6	30'	90%	0° W	-G	Jupiter
4th	Sco	16h 06m	-16° 0'	-11.2	30'	83%	0° W	-G	Jupiter, Antares
5th	Oph	16h 57m	-18° 18'	-10.9	30'	76%	0° W	-G	Antares
6th	Sgr	17h 47m	-19° 45'	-10.5	30'	67%	0° W	-G	
7th	Sgr	18h 38m	-20° 18'	-10.1	30'	58%	0° W	LQ	Mars, Saturn
8th	Sgr	19h 28m	-19° 57'	-9.7	30'	48%	0° W	LQ	Mars, Saturn
9th	Cap	20h 18m	-18° 42'	-9.3	30'	39%	0° W	LQ	
10th	Cap	21h 08m	-16° 39'	-8.8	30'	30%	0° W	-Cr	

Mercury and Venus

| Mercury | Mercury | Venus |
| 3rd | 7th | 5th |

Mercury

Date	Con	R.A.	Dec	Mag	Diam	Ill.	Elong	AM/PM	Close To
1st	Psc	00h 40m	7° 27'	5.2	11''	0%	3° E	NV	
3rd	Psc	00h 35m	6° 27'	4.9	11''	1%	4° W	NV	
5th	Psc	00h 30m	5° 24'	4.2	11''	2%	7° W	NV	
7th	Psc	00h 25m	4° 21'	3.5	11''	5%	10° W	NV	
9th	Psc	00h 22m	3° 21'	2.8	11''	7%	13° W	NV	

Venus

Date	Con	R.A.	Dec	Mag	Diam	Ill	Elong	AM/PM	Close To
1st	Ari	01h 58m	11° 36'	-3.9	11"	94%	20° E	PM	Uranus
3rd	Ari	02h 08m	12° 30'	-3.9	11"	94%	20° E	PM	
5th	Ari	02h 17m	13° 24'	-3.9	11"	94%	21° E	PM	
7th	Ari	02h 27m	14° 18'	-3.9	11"	93%	21° E	PM	
9th	Ari	02h 36m	15° 9'	-3.9	11"	93%	22° E	PM	

Mars and the Outer Planets

Mars
5th

Jupiter
5th

Saturn
5th

Mars

Date	Con	R.A.	Dec	Mag	Diam	Ill	Elong	AM/PM	Close To
1st	Sgr	18h 36m	-23° 30'	0.3	9"	88%	93° W	AM	Saturn
5th	Sgr	18h 46m	-23° 30'	0.2	9"	88%	95° W	AM	Saturn
10th	Sgr	18h 58m	-23° 24'	0.1	9"	88%	97° W	AM	Saturn

The Outer Planets

Planet	Date	Con	R.A.	Dec	Mag	Diam	Elong	AM/PM	Close To
Jupiter	5th	Lib	15h 20m	-15° 3'	-2.4	43"	144° W	AM	
Saturn	5th	Sgr	18h 39m	-22° 9'	0.5	17"	97° W	AM	Mars
Uranus	5th	Psc	01h 44m	10° 9'	5.9	3"	12° E	NV	
Neptune	5th	Aqr	23h 07m	-6° 42'	8.0	2"	31° W	AM	

Events

Date	Time (UT)	Event
1st	08:35	The just-past full Moon appears north of the bright star Spica. (Virgo, all night.)
	17:47	Mercury is at inferior conjunction with the Sun. (Not visible.)
2nd	11:49	Mars appears 1.3° south of Saturn. (Morning sky.)
3rd	20:59	The waning gibbous Moon appears north of Jupiter. (Morning sky.)
4th	21:13	The waning gibbous Moon appears north of the bright star Antares. (Scorpius, morning sky.)
7th	14:02	The nearly last quarter Moon appears north of Saturn. (Morning sky.)
	18:06	The nearly last quarter Moon appears north of Mars. (Morning sky.)
8th	07:18	Last Quarter Moon. (Morning sky.)
	14:33	The last quarter Moon appears to the north of dwarf planet Pluto. (Morning sky.)

Planet Locations – April 5th

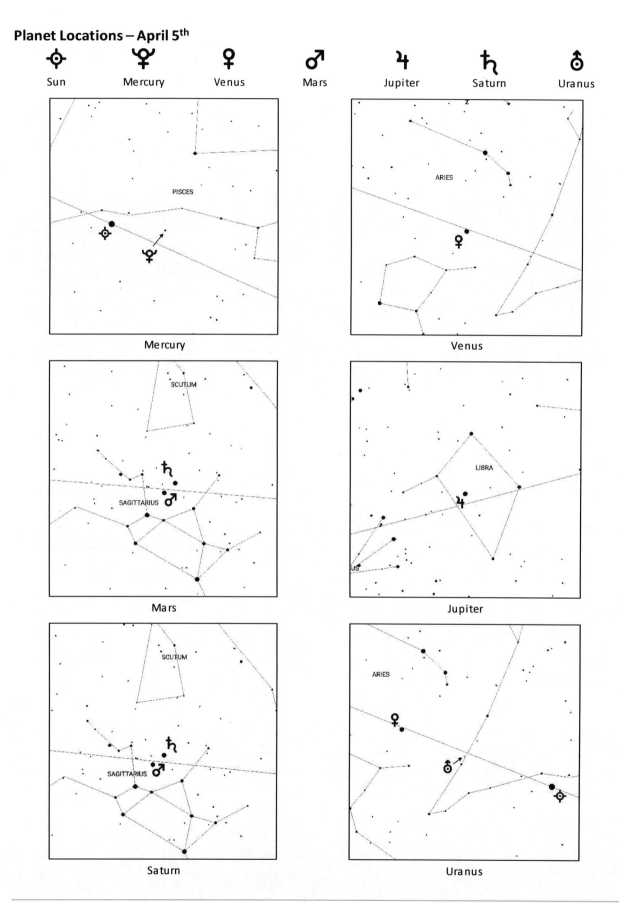

The Moon

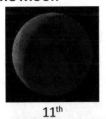    

| 11th | 13th | 15th | 17th | 19th |

Date	Con	R.A.	Dec	Mag	Diam	Ill.	Elong	Phase	Close To
11th	Cap	21h 57m	-13° 48'	-8.2	30'	21%	0° W	-Cr	
12th	Aqr	22h 46m	-10° 18'	-7.6	31'	14%	0° W	-Cr	Neptune
13th	Aqr	23h 34m	-6° 15'	-6.8	31'	7%	0° W	NM	
14th	Psc	00h 24m	-1° 48'	-6.0	32'	3%	0° W	NM	Mercury
15th	Psc	01h 14m	2° 51'	-5.1	32'	1%	0° W	NM	
16th	Psc	02h 05m	7° 27'	-5.0	32'	0%	0° W	NM	Uranus
17th	Ari	02h 59m	11° 45'	-5.9	33'	3%	0° W	NM	Venus, Pleiades
18th	Tau	03h 55m	15° 27'	-6.8	33'	8%	0° W	NM	Venus, Pleiades, Hyades
19th	Tau	04h 54m	18° 18'	-7.7	33'	15%	0° W	+Cr	Hyades, Aldebaran
20th	Ori	05h 54m	20° 0'	-8.4	33'	24%	0° W	+Cr	

Mercury and Venus

| Mercury | Mercury | Venus |
| 13th | 17th | 15th |

Mercury

Date	Con	R.A.	Dec	Mag	Diam	Ill.	Elong	AM/PM	Close To
11th	Psc	00h 19m	2° 33'	2.3	11''	11%	16° W	AM	
13th	Psc	00h 18m	1° 51'	1.9	11''	15%	19° W	AM	
15th	Psc	00h 19m	1° 21'	1.6	10''	18%	21° W	AM	
17th	Psc	00h 20m	1° 0'	1.3	10''	22%	23° W	AM	
19th	Psc	00h 22m	0° 51'	1.1	10''	26%	24° W	AM	

Venus

Date	Con	R.A.	Dec	Mag	Diam	Ill	Elong	AM/PM	Close To
11th	Ari	02h 46m	15° 57'	-3.9	11"	92%	22° E	PM	
13th	Ari	02h 55m	16° 48'	-3.9	11"	92%	23° E	PM	
15th	Ari	03h 05m	17° 33'	-3.9	11"	92%	23° E	PM	
17th	Ari	03h 15m	18° 18'	-3.9	11"	91%	24° E	PM	Moon, Pleiades
19th	Tau	03h 25m	19° 0'	-3.9	11"	91%	24° E	PM	Pleiades

Mars and the Outer Planets

Mars
15th

Jupiter
15th

Saturn
15th

Mars

Date	Con	R.A.	Dec	Mag	Diam	Ill	Elong	AM/PM	Close To
10th	Sgr	18h 58m	-23° 24'	0.1	9"	88%	97° W	AM	Saturn
15th	Sgr	19h 09m	-23° 15'	0.0	10"	88%	100° W	AM	Saturn
20th	Sgr	19h 20m	-23° 6'	-0.1	10"	88%	102° W	AM	

The Outer Planets

Planet	Date	Con	R.A.	Dec	Mag	Diam	Elong	AM/PM	Close To
Jupiter	15th	Lib	15h 16m	-16° 48'	-2.4	44"	154° W	AM	
Saturn	15th	Sgr	18h 40m	-22° 15'	0.5	17"	106° W	AM	Mars
Uranus	15th	Psc	01h 46m	10° 24'	5.9	3"	3° E	NV	Moon
Neptune	15th	Aqr	23h 08m	-6° 36'	7.9	2"	40° W	AM	

Events

Date	Time (UT)	Event
12th	N/A	Good opportunity to see Earthshine on the waning crescent Moon. (Morning sky.)
	22:25	The waning crescent Moon appears to the south of Neptune. (Morning sky.)
14th	03:50	Mercury is stationary prior to resuming prograde motion. (Morning sky.)
	08:21	The waning crescent Moon appears south of Mercury. (Morning sky.)
16th	01:58	New Moon. (Not visible.)
18th	00:17	Saturn is stationary prior to beginning retrograde motion. (Morning sky.)
	06:51	The waxing crescent Moon appears south of the Pleiades star cluster. (Taurus, evening sky.)
	N/A	Good opportunity to see Earthshine on the waxing crescent Moon. (Evening sky.)
19th	04:09	The waxing crescent Moon appears north of the bright star Aldebaran. (Taurus, evening sky.)

Planet Locations – April 15th

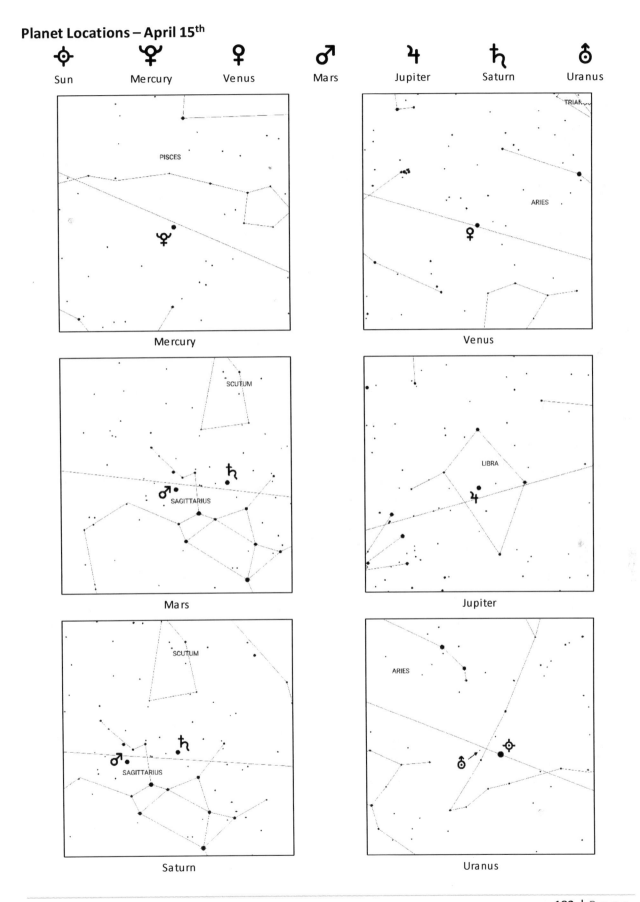

April 21st – 30th

The Moon

21st 23rd 25th 27th 29th

Date	Con	R.A.	Dec	Mag	Diam	Ill.	Elong	Phase	Close To
21st	Gem	06h 55m	20° 23'	-9.0	33'	34%	0° W	+Cr	
22nd	Cnc	07h 55m	19° 33'	-9.6	32'	46%	0° W	FQ	Praesepe
23rd	Cnc	08h 54m	17° 27'	-10.1	32'	57%	0° W	FQ	Praesepe
24th	Leo	09h 51m	14° 18'	-10.5	32'	68%	0° W	+G	Regulus
25th	Leo	10h 45m	10° 24'	-11.0	32'	78%	0° W	+G	Regulus
26th	Vir	11h 38m	6° 0'	-11.3	31'	86%	0° W	+G	
27th	Vir	12h 29m	1° 21'	-11.7	31'	93%	0° W	+G	
28th	Vir	13h 19m	-3° 18'	-12.1	31'	97%	0° W	FM	Spica
29th	Vir	14h 09m	-7° 42'	-12.4	30'	100%	0° W	FM	Spica
30th	Lib	14h 58m	-11° 42'	-12.5	30'	100%	0° W	FM	Jupiter

Mercury and Venus

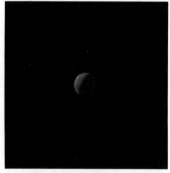 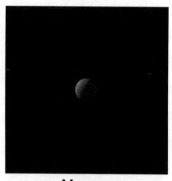 

Mercury   Mercury   Venus
23rd 27th 25th

Mercury

Date	Con	R.A.	Dec	Mag	Diam	Ill.	Elong	AM/PM	Close To
21st	Psc	00h 26m	0° 48'	1.0	9''	30%	25° W	AM	
23rd	Cet	00h 31m	1° 0'	0.8	9''	33%	26° W	AM	
25th	Cet	00h 36m	1° 18'	0.7	9''	37%	27° W	AM	
27th	Cet	00h 42m	1° 42'	0.6	8''	40%	27° W	AM	
29th	Cet	00h 49m	2° 15'	0.5	8''	43%	27° W	AM	

Venus

Date	Con	R.A.	Dec	Mag	Diam	Ill	Elong	AM/PM	Close To
21st	Tau	03h 35m	19° 42'	-3.9	11"	91%	25° E	PM	Pleiades
23rd	Tau	03h 45m	20° 21'	-3.9	11"	90%	25° E	PM	Pleiades
25th	Tau	03h 55m	20° 57'	-3.9	11"	90%	26° E	PM	Pleiades, Hyades
27th	Tau	04h 05m	21° 30'	-3.9	11"	89%	26° E	PM	Pleiades, Hyades
29th	Tau	04h 15m	22° 3'	-3.9	11"	89%	27° E	PM	Pleiades, Hyades, Aldebaran

Mars and the Outer Planets

Mars
25th

Jupiter
25th

Saturn
25th

Mars

Date	Con	R.A.	Dec	Mag	Diam	Ill	Elong	AM/PM	Close To
20th	Sgr	19h 20m	-23° 6'	-0.1	10"	88%	102° W	AM	
25th	Sgr	19h 31m	-22° 54'	-0.2	11"	88%	104° W	AM	
30th	Sgr	19h 41m	-22° 42'	-0.4	11"	88%	107° W	AM	

The Outer Planets

Planet	Date	Con	R.A.	Dec	Mag	Diam	Elong	AM/PM	Close To
Jupiter	25th	Lib	15h 12m	-16° 30'	-2.5	44"	165° W	AM	
Saturn	25th	Sgr	18h 39m	-22° 15'	0.4	17"	116° W	AM	
Uranus	25th	Ari	01h 48m	10° 36'	5.9	3"	6° W	NV	
Neptune	25th	Aqr	23h 09m	-6° 27'	7.9	2"	50° W	AM	

Events

Date	Time (UT)	Event
22nd	N/A	The Lyrid meteors are at their peak. (ZHR: 18. All night but best in the morning sky.)
	21:46	First Quarter Moon. (Evening sky.)
23rd	06:22	The just-past first quarter Moon appears to the south of the Praesepe star cluster. (Cancer, evening sky.)
	22:30	Venus appears 3.6° south of the Pleiades star cluster. (Taurus, evening sky.)
24th	19:30	The waxing gibbous Moon appears north of the bright star Regulus. (Leo, evening sky.)
28th	13:37	The waxing gibbous Moon appears north of the bright star Spica. (Virgo, evening sky.)
29th	18:12	Mercury is at greatest western elongation from the Sun. (Morning sky.)
30th	00:59	Full Moon. (All night.)
	15:51	The full Moon appears to the north of Jupiter. (All night.)

Planet Locations – April 25th

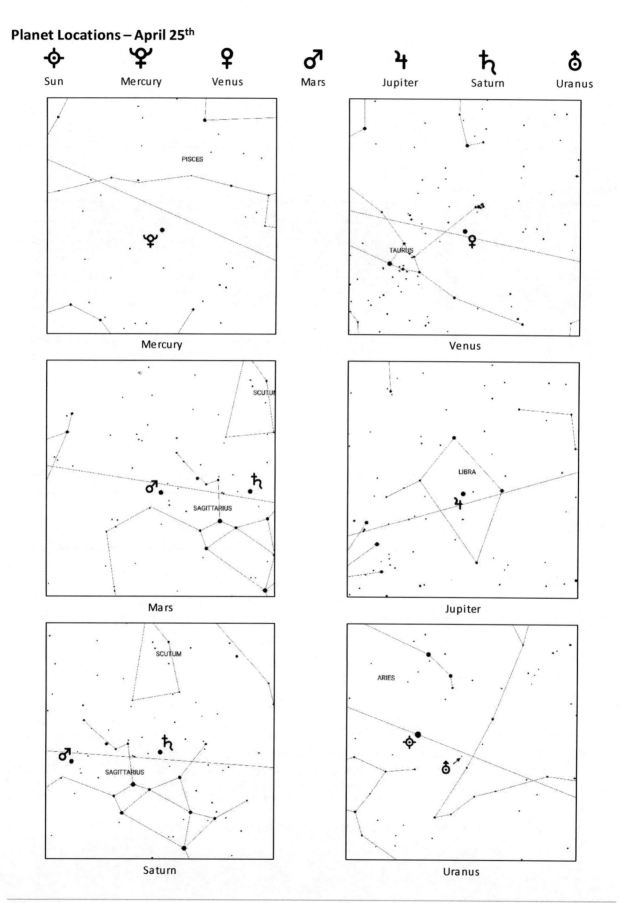

The Moon

| | 1ˢᵗ | | 3ʳᵈ | | 5ᵗʰ | | 7ᵗʰ | | 9ᵗʰ |

Date	Con	R.A.	Dec	Mag	Diam	Ill.	Elong	Phase		Close To
1st	Lib	15h 48m	-15° 6'	-12.1	30'	98%	162° W	FM		Jupiter, Antares
2nd	Oph	16h 39m	-17° 45'	-11.8	30'	94%	151° W	-G		Antares
3rd	Oph	17h 30m	-19° 33'	-11.5	30'	88%	140° W	-G		
4th	Sgr	18h 21m	-20° 27'	-11.1	29'	82%	129° W	-G		Saturn
5th	Sgr	19h 12m	-20° 24'	-10.8	29'	74%	118° W	-G		Mars, Saturn
6th	Sgr	20h 02m	-19° 30'	-10.4	29'	65%	107° W	LQ		
7th	Cap	20h 52m	-17° 42'	-10.1	30'	56%	97° W	LQ		
8th	Cap	21h 40m	-15° 6'	-9.6	30'	46%	86° W	LQ		
9th	Aqr	22h 29m	-11° 51'	-9.2	30'	37%	74° W	-Cr		
10th	Aqr	23h 17m	-8° 0'	-8.6	31'	27%	63° W	-Cr		Neptune

Mercury and Venus

Mercury	Mercury	Venus
3ʳᵈ	7ᵗʰ	5ᵗʰ

Mercury

Date	Con	R.A.	Dec	Mag	Diam	Ill.	Elong	AM/PM		Close To
1st	Psc	00h 57m	2° 57'	0.4	8''	47%	27° W	AM		
3rd	Psc	01h 05m	3° 45'	0.3	7''	50%	27° W	AM		
5th	Psc	01h 14m	4° 36'	0.2	7''	53%	26° W	AM		
7th	Psc	01h 24m	5° 33'	0.1	7''	56%	26° W	AM		
9th	Psc	01h 34m	6° 36'	0.0	7''	59%	25° W	AM		Uranus

Venus

Date	Con	R.A.	Dec	Mag	Diam	Ill	Elong	AM/PM	Close To
1st	Tau	04h 26m	22° 33'	-3.9	12''	88%	27° E	PM	Pleiades, Hyades, Aldebaran
3rd	Tau	04h 36m	23° 0'	-3.9	12''	88%	28° E	PM	Hyades, Aldebaran
5th	Tau	04h 46m	23° 23'	-3.9	12''	87%	28° E	PM	Hyades, Aldebaran
7th	Tau	04h 57m	23° 48'	-3.9	12''	87%	29° E	PM	Hyades, Aldebaran
9th	Tau	05h 07m	24° 6'	-3.9	12''	87%	29° E	PM	Aldebaran

Mars and the Outer Planets

Mars
5th

Jupiter
5th

Saturn
5th

Mars

Date	Con	R.A.	Dec	Mag	Diam	Ill	Elong	AM/PM	Close To
1st	Sgr	19h 43m	-22° 39'	-0.4	11''	88%	107° W	AM	
5th	Sgr	19h 51m	-22° 30'	-0.5	12''	89%	109° W	AM	Moon
10th	Sgr	00h 00m	-22° 18'	-0.6	12''	89%	112° W	AM	

The Outer Planets

Planet	Date	Con	R.A.	Dec	Mag	Diam	Elong	AM/PM	Close To
Jupiter	5th	Lib	15h 07m	-16° 12'	-2.5	45''	176° W	AM	
Saturn	5th	Sgr	18h 39m	-22° 15'	0.3	18''	126° W	AM	Moon
Uranus	5th	Ari	01h 50m	10° 48'	5.9	3''	16° W	AM	
Neptune	5th	Aqr	23h 10m	-6° 24'	7.9	2''	59° W	AM	

Events

Date	Time (UT)	Event
2nd	09:08	The waning gibbous Moon appears to the north of the bright star Antares. (Scorpius, morning sky.)
3rd	12:24	Venus appears 6.5° north of the bright star Aldebaran. (Taurus, evening sky.)
4th	18:55	The waning gibbous Moon appears to the north of Saturn. (Morning sky.)
5th	19:35	The waning gibbous Moon appears to the north of dwarf planet Pluto. (Morning sky.)
	N/A	The Eta Aquariids meteors reach their peak. (ZHR: 65. All night but best seen in the morning sky.)
	09:05	The waning gibbous Moon appears to the north of Mars. (Morning sky.)
8th	02:09	Last Quarter Moon. (Morning sky.)
9th	08:23	Jupiter is at opposition. (All night.)
10th	09:38	The waning crescent Moon appears to the south of Neptune. (Morning sky.)

Planet Locations – May 5th

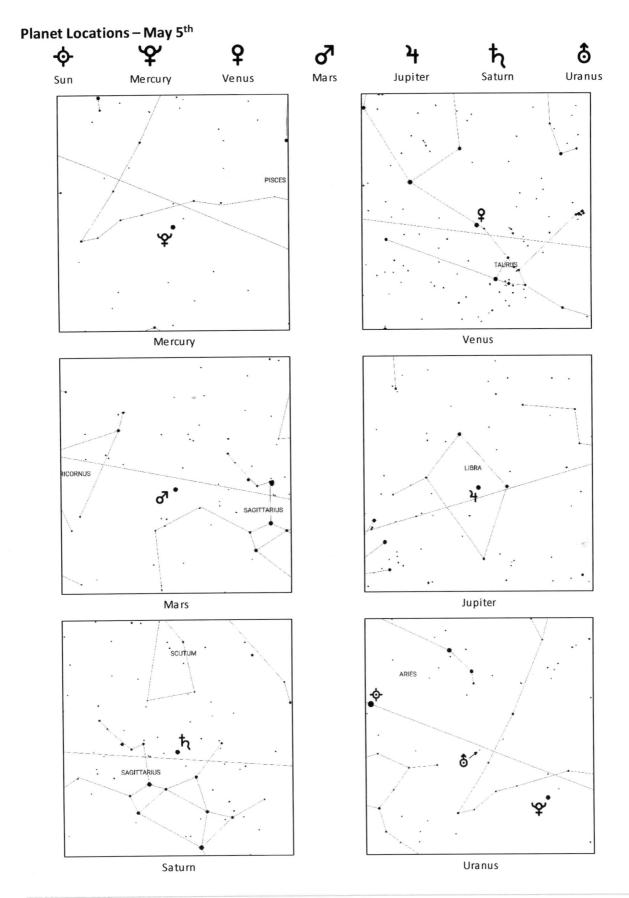

May 11th – 20th

The Moon

| 11th | 13th | 15th | 17th | 19th |

Date	Con	R.A.	Dec	Mag	Diam	Ill.	Elong	Phase	Close To
11th	Psc	00h 05m	-3° 42'	-8.0	31'	19%	51° W	-Cr	
12th	Cet	00h 54m	0° 54'	-7.3	32'	11%	39° W	NM	
13th	Psc	01h 45m	5° 36'	-6.5	32'	5%	27° W	NM	Mercury, Uranus
14th	Cet	02h 38m	10° 9'	-5.5	33'	2%	14° W	NM	Mercury
15th	Tau	03h 34m	14° 45'	-4.7	33'	0%	5° E	NM	Pleiades
16th	Tau	04h 33m	17° 33'	-5.5	33'	2%	14° E	NM	Pleiades, Hyades, Aldebaran
17th	Tau	05h 35m	19° 45'	-6.5	33'	6%	28° E	NM	Venus
18th	Gem	06h 37m	20° 38'	-7.4	33'	13%	41° E	+Cr	Venus
19th	Gem	07h 39m	20° 6'	-8.2	33'	21%	55° E	+Cr	
20th	Cnc	08h 40m	18° 18'	-8.9	33'	32%	68° E	+Cr	Praesepe

Mercury and Venus

| Mercury | Mercury | Venus |
| 13th | 17th | 15th |

Mercury

Date	Con	R.A.	Dec	Mag	Diam	Ill.	Elong	AM/PM	Close To
11th	Psc	01h 44m	7° 45'	0.0	7''	62%	24° W	AM	Uranus
13th	Psc	01h 55m	8° 54'	-0.1	6''	66%	23° W	AM	Moon, Uranus
15th	Cet	02h 07m	10° 9'	-0.3	6''	69%	22° W	AM	Uranus
17th	Ari	02h 19m	11° 24'	-0.4	6''	72%	20° W	AM	
19th	Ari	02h 32m	12° 45'	-0.5	6''	76%	19° W	AM	

Venus

Date	Con	R.A.	Dec	Mag	Diam	Ill	Elong	AM/PM		Close To
11th	Tau	05h 18m	24° 21'	-3.9	12''	86%	30° E	PM		
13th	Tau	05h 28m	24° 36'	-4.0	12''	85%	30° E	PM		
15th	Tau	05h 39m	24° 48'	-4.0	12''	85%	31° E	PM		
17th	Tau	05h 49m	24° 53'	-4.0	12''	84%	31° E	PM		Moon
19th	Tau	06h 00m	25° 0'	-4.0	12''	84%	32° E	PM		

Mars and the Outer Planets

Mars
15th

Jupiter
15th

Saturn
15th

Mars

Date	Con	R.A.	Dec	Mag	Diam	Ill	Elong	AM/PM		Close To
10th	Sgr	00h 00m	-22° 18'	-0.6	12''	89%	112° W	AM		
15th	Cap	20h 09m	-22° 6'	-0.7	13''	89%	115° W	AM		
20th	Cap	20h 18m	-21° 57'	-0.9	14''	90%	118° W	AM		

The Outer Planets

Planet	Date	Con	R.A.	Dec	Mag	Diam	Elong	AM/PM		Close To
Jupiter	15th	Lib	15h 02m	-15° 51'	-2.5	45''	173° E	PM		
Saturn	15th	Sgr	18h 37m	-22° 15'	0.3	18''	136° W	AM		
Uranus	15th	Ari	01h 52m	10° 57'	5.9	3''	25° W	AM		Mercury
Neptune	15th	Aqr	23h 11m	-6° 18'	7.9	2''	68° W	AM		

Events

Date	Time (UT)	Event
11th	N/A	Good opportunity to see Earthshine on the waning crescent Moon. (Morning sky.)
12th	21:00	Mercury appears 2.4° south of Uranus. (Morning sky.)
13th	17:03	The waning crescent Moon appears south of Uranus. (Morning sky.)
	19:00	The waning crescent Moon appears south of Mercury. (Morning sky.)
15th	11:48	New Moon. (Not visible.)
17th	20:00	The waxing crescent Moon appears to the south of Venus. (Evening sky.)
18th	N/A	Good opportunity to see Earthshine on the waxing crescent Moon. (Evening sky.)
20th	10:14	The waxing crescent Moon appears to the south of the Praesepe star cluster. (Cancer, evening sky.)

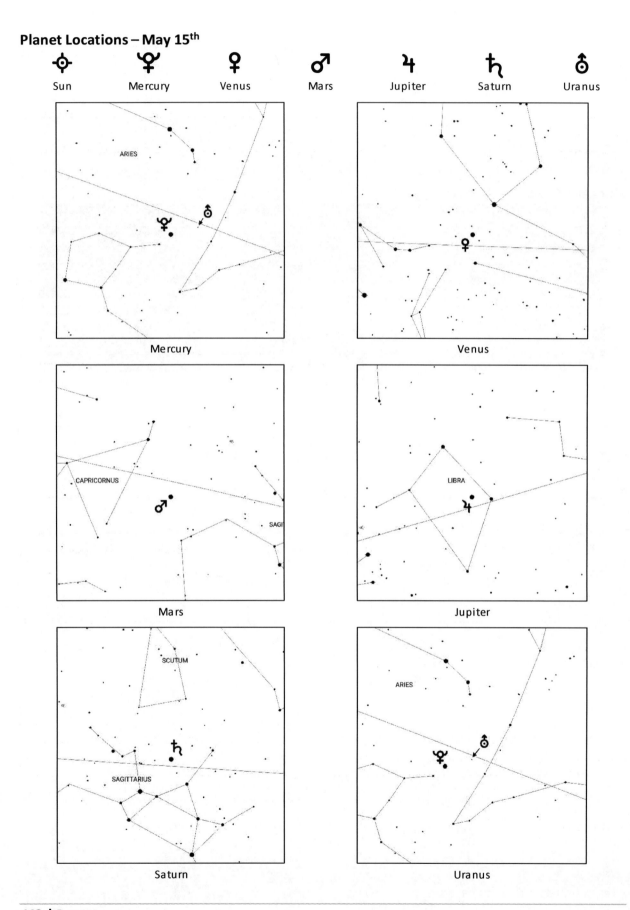

| Sun | Mercury | Venus | Mars | Jupiter | Saturn | Uranus |

Mercury

Venus

Mars

Jupiter

Saturn

Uranus

The Moon

| 21st | 23rd | 25th | 27th | 29th | 31st |

Date	Con	R.A.	Dec	Mag	Diam	Ill.	Elong	Phase	Close To
21st	Leo	09h 38m	15° 21'	-9.5	32'	43%	81° E	FQ	Regulus
22nd	Leo	10h 34m	11° 36'	-10.0	32'	54%	94° E	FQ	Regulus
23rd	Leo	11h 26m	7° 15'	-10.4	31'	65%	107° E	FQ	
24th	Vir	12h 17m	2° 39'	-10.8	31'	75%	120° E	+G	
25th	Vir	13h 06m	-2° 0'	-11.2	31'	83%	132° E	+G	Spica
26th	Vir	13h 55m	-6° 27'	-11.6	30'	90%	144° E	+G	Spica
27th	Lib	14h 44m	-10° 36'	-11.9	30'	95%	155° E	FM	Jupiter
28th	Lib	15h 34m	-14° 12'	-12.3	30'	99%	167° E	FM	Jupiter
29th	Oph	16h 24m	-17° 6'	-12.5	30'	100%	175° E	FM	Antares
30th	Oph	17h 14m	-19° 12'	-12.3	29'	99%	169° W	FM	Antares
31st	Sgr	18h 05m	-20° 24'	-12.0	29'	97%	159° W	FM	Saturn

Mercury and Venus

Mercury
23rd

Mercury
27th

Venus
25th

Mercury

Date	Con	R.A.	Dec	Mag	Diam	Ill.	Elong	AM/PM	Close To
21st	Ari	02h 46m	14° 3'	-0.7	6''	79%	17° W	AM	
23rd	Ari	03h 00m	15° 15'	-0.8	6''	83%	15° W	NV	
25th	Ari	03h 15m	16° 45'	-1.0	5''	87%	13° W	NV	Pleiades
27th	Tau	03h 31m	18° 6'	-1.2	5''	90%	11° W	NV	Pleiades
29th	Tau	03h 48m	19° 21'	-1.4	5''	94%	9° W	NV	Pleiades, Hyades
31st	Tau	04h 05m	20° 33'	-1.6	5''	96%	7° W	NV	Pleiades, Hyades, Aldebaran

Venus

Date	Con	R.A.	Dec	Mag	Diam	Ill	Elong	AM/PM	Close To
21st	Gem	06h 10m	25° 3'	-4.0	13''	83%	32° E	PM	
23rd	Gem	06h 21m	25° 3'	-4.0	13''	83%	32° E	PM	
25th	Gem	06h 31m	25° 0'	-4.0	13''	82%	33° E	PM	
27th	Gem	06h 42m	24° 53'	-4.0	13''	82%	33° E	PM	
29th	Gem	06h 52m	24° 45'	-4.0	13''	81%	34° E	PM	
31st	Gem	07h 03m	24° 33'	-4.0	13''	80%	34° E	PM	

Mars and the Outer Planets

Mars
25th

Jupiter
25th

Saturn
25th

Mars

Date	Con	R.A.	Dec	Mag	Diam	Ill	Elong	AM/PM	Close To
20th	Cap	20h 18m	-21° 57'	-0.9	14''	90%	118° W	AM	
25th	Cap	20h 25m	-21° 48'	-1.0	14''	90%	121° W	AM	
30th	Cap	20h 32m	-21° 42'	-1.2	15''	91%	124° W	AM	

The Outer Planets

Planet	Date	Con	R.A.	Dec	Mag	Diam	Elong	AM/PM	Close To
Jupiter	25th	Lib	14h 57m	-15° 30'	-2.5	44''	162° E	PM	
Saturn	25th	Sgr	18h 35m	-22° 18'	0.2	18''	146° W	AM	
Uranus	25th	Ari	01h 54m	11° 9'	5.9	3''	34° W	AM	
Neptune	25th	Aqr	23h 11m	-6° 15'	7.9	2''	78° W	AM	

Events

Date	Time (UT)	Event
22nd	02:20	The almost first quarter Moon appears north of the bright star Regulus. (Leo, evening sky.)
	03:50	First Quarter Moon. (Evening sky.)
25th	21:07	The waxing gibbous Moon appears north of the bright star Spica. (Virgo, evening sky.)
27th	15:50	The waxing gibbous Moon appears north of Jupiter. (Evening sky.)
29th	13:51	The almost full Moon appears to the north of the bright star Antares. (Scorpius, all night.)
	14:20	Full Moon. (All night.)

Planet Locations – May 25th

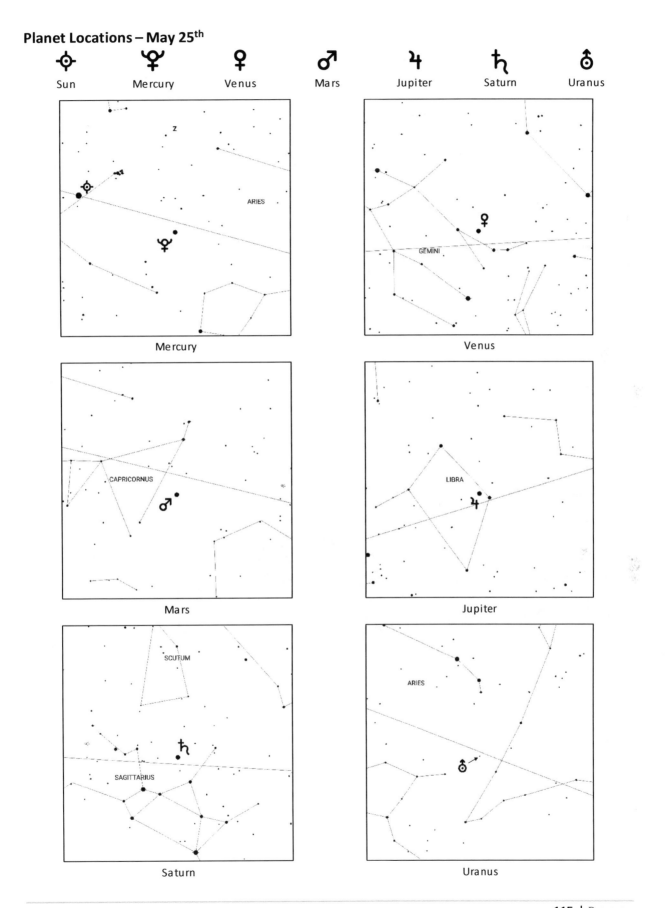

| Sun | Mercury | Venus | Mars | Jupiter | Saturn | Uranus |

Mercury

Venus

Mars

Jupiter

Saturn

Uranus

June 1st to 10th

The Moon

1st

3rd

5th

7th

9th

Date	Con	R.A.	Dec	Mag	Diam	Ill.	Elong	Phase	Close To
1st	Sgr	18h 57m	-20° 42'	-11.7	29'	92%	148° W	-G	Saturn
2nd	Sgr	19h 47m	-20° 3'	-11.4	29'	87%	137° W	-G	Mars
3rd	Cap	20h 37m	-18° 33'	-11.0	29'	80%	126° W	-G	Mars
4th	Cap	21h 26m	-16° 12'	-10.7	29'	71%	115° W	-G	Mars
5th	Aqr	22h 13m	-13° 12'	-10.3	30'	62%	104° W	LQ	
6th	Aqr	23h 00m	-9° 36'	-9.9	30'	53%	93° W	LQ	Neptune
7th	Aqr	23h 48m	-5° 27'	-9.5	31'	43%	82° W	LQ	
8th	Cet	00h 36m	-1° 3'	-9.0	31'	33%	70° W	-Cr	
9th	Psc	01h 25m	3° 36'	-8.4	32'	24%	58° W	-Cr	
10th	Cet	02h 16m	8° 12'	-7.7	32'	15%	45° W	-Cr	Uranus

Mercury and Venus

Mercury
3rd

Mercury
7th

Venus
5th

Mercury

Date	Con	R.A.	Dec	Mag	Diam	Ill.	Elong	AM/PM	Close To
1st	Tau	04h 14m	21° 8'	-1.7	5''	98%	6° W	NV	Pleiades, Hyades, Aldebaran
3rd	Tau	04h 32m	22° 12'	-2.0	5''	99%	3° W	NV	Hyades, Aldebaran
5th	Tau	04h 51m	23° 6'	-2.3	5''	100%	1° W	NV	Hyades, Aldebaran
7th	Tau	05h 10m	23° 53'	-2.1	5''	100%	2° E	NV	Aldebaran
9th	Tau	05h 29m	24° 30'	-1.9	5''	98%	4° E	NV	

Venus

Date	Con	R.A.	Dec	Mag	Diam	Ill	Elong	AM/PM	Close To
1st	Gem	07h 08m	24° 27'	-4.0	13''	80%	35° E	PM	
3rd	Gem	07h 18m	24° 8'	-4.0	13''	79%	35° E	PM	
5th	Gem	07h 28m	23° 51'	-4.0	13''	79%	35° E	PM	
7th	Gem	07h 39m	23° 30'	-4.0	14''	78%	36° E	PM	
9th	Gem	07h 48m	23° 6'	-4.0	14''	78%	36° E	PM	

Mars and the Outer Planets

Mars
5th

Jupiter
5th

Saturn
5th

Mars

Date	Con	R.A.	Dec	Mag	Diam	Ill	Elong	AM/PM	Close To
1st	Cap	20h 35m	-21° 42'	-1.2	15''	91%	125° W	AM	
5th	Cap	20h 39m	-21° 42'	-1.4	16''	92%	128° W	AM	
10th	Cap	20h 44m	-21° 45'	-1.5	17''	93%	132° W	AM	

The Outer Planets

Planet	Date	Con	R.A.	Dec	Mag	Diam	Elong	AM/PM	Close To
Jupiter	5th	Lib	14h 52m	-15° 12'	-2.4	44''	150° E	PM	
Saturn	5th	Sgr	18h 32m	-22° 21'	0.1	18''	157° W	AM	
Uranus	5th	Ari	01h 56m	11° 21'	5.9	3''	44° W	AM	
Neptune	5th	Aqr	23h 12m	-6° 12'	7.9	2''	88° W	AM	

Events

Date	Time (UT)	Event
1st	00:20	The waning gibbous Moon appears north of Saturn. (Morning sky)
2nd	04:04	The waning gibbous Moon appears north of the dwarf planet Pluto. (Morning sky.)
3rd	13:11	The waning gibbous Moon appears north of Mars. (Morning sky.)
6th	01:49	Mercury is at superior conjunction with the Sun. (Not visible.)
	17:46	The almost last quarter Moon appears south of Neptune. (Morning sky.)
	18:32	Last Quarter Moon. (Morning sky.)
7th	N/A	The Arietid meteors are at their peak. (ZHR: 54. All night but best in the morning sky.)
10th	01:33	The waning crescent Moon appears south of Uranus. (Morning sky.)
	N/A	Good opportunity to see Earthshine on the waning crescent Moon. (Morning sky.)

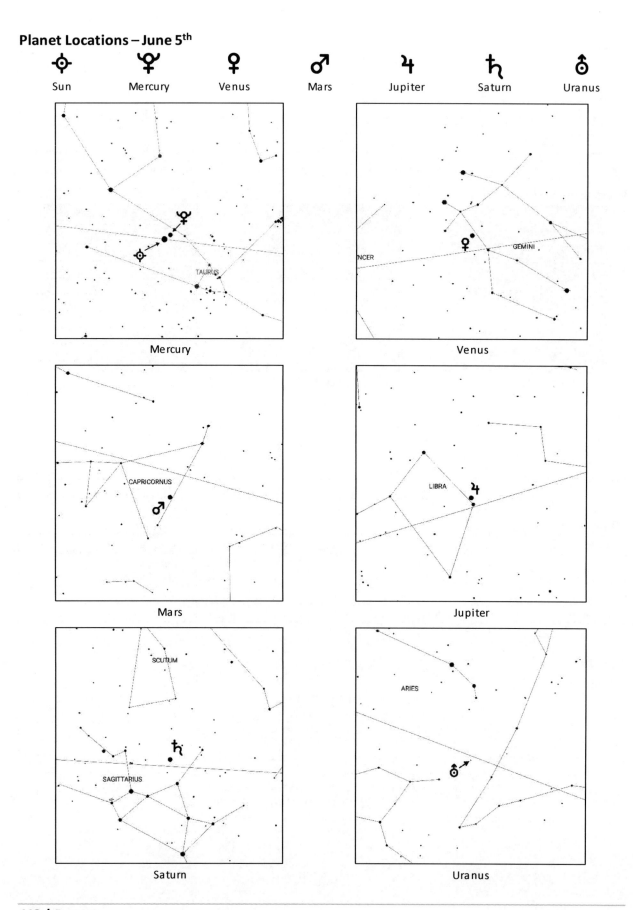

The Moon

| | 11th | 13th | 15th | 17th | 19th |

Date	Con	R.A.	Dec	Mag	Diam	Ill.	Elong	Phase	Close To
11th	Ari	03h 10m	12° 30'	-6.9	33'	8%	32° W	NM	Pleiades
12th	Tau	04h 08m	16° 15'	-5.9	33'	3%	19° W	NM	Pleiades, Hyades, Aldebaran
13th	Tau	05h 09m	19° 3'	-4.8	34'	0%	6° W	NM	Hyades, Aldebaran
14th	Ori	06h 12m	20° 33'	-5.2	34'	1%	10° E	NM	Mercury
15th	Gem	07h 17m	20° 36'	-6.3	34'	4%	24° E	NM	
16th	Cnc	08h 20m	19° 12'	-7.2	34'	10%	38° E	NM	Venus, Praesepe
17th	Leo	09h 21m	16° 30'	-8.0	33'	19%	51° E	+Cr	Praesepe, Regulus
18th	Leo	10h 19m	12° 51'	-8.7	33'	29%	65° E	+Cr	Regulus
19th	Leo	11h 14m	8° 33'	-9.3	32'	40%	78° E	FQ	
20th	Vir	12h 06m	3° 57'	-9.8	32'	51%	91° E	FQ	

Mercury and Venus

Mercury	Mercury	Venus
13th	17th	15th

Mercury

Date	Con	R.A.	Dec	Mag	Diam	Ill.	Elong	AM/PM	Close To
11th	Tau	05h 48m	24° 53'	-1.6	5''	96%	7° E	NV	
13th	Gem	06h 06m	25° 6'	-1.4	5''	93%	9° E	NV	
15th	Gem	06h 25m	25° 8'	-1.2	5''	90%	11° E	NV	
17th	Gem	06h 42m	25° 0'	-1.0	5''	86%	13° E	NV	
19th	Gem	06h 59m	24° 45'	-0.8	6''	82%	15° E	NV	

Venus

Date	Con	R.A.	Dec	Mag	Diam	Ill	Elong	AM/PM	Close To
11th	Gem	07h 59m	22° 42'	-4.0	14''	77%	37° E	PM	
13th	Cnc	08h 09m	22° 12'	-4.0	14''	76%	37° E	PM	
15th	Cnc	08h 18m	21° 42'	-4.0	14''	76%	38° E	PM	
17th	Cnc	08h 28m	21° 8'	-4.0	14''	75%	38° E	PM	Praesepe
19th	Cnc	08h 38m	20° 33'	-4.0	15''	74%	38° E	PM	Praesepe

Mars and the Outer Planets

Mars
15th

Jupiter
15th

Saturn
15th

Mars

Date	Con	R.A.	Dec	Mag	Diam	Ill	Elong	AM/PM	Close To
10th	Cap	20h 44m	-21° 45'	-1.5	17''	93%	132° W	AM	
15th	Cap	20h 46m	-21° 51'	-1.7	18''	94%	136° W	AM	
20th	Cap	20h 50m	-22° 3'	-1.8	19''	95%	140° W	AM	

The Outer Planets

Planet	Date	Con	R.A.	Dec	Mag	Diam	Elong	AM/PM	Close To
Jupiter	15th	Lib	14h 49m	-15° 0'	-2.4	43''	140° E	PM	
Saturn	15th	Sgr	18h 29m	-22° 24'	0.1	18''	168° W	AM	
Uranus	15th	Ari	01h 58m	11° 30'	5.9	3''	53° W	AM	
Neptune	15th	Aqr	23h 12m	-6° 12'	7.0	2''	98° W	AM	

Events

Date	Time (UT)	Event
13th	13:55	New Moon. (Not visible.)
16th	12:09	The waxing crescent Moon appears south of Venus. (Evening sky.)
	21:18	The waxing crescent Moon appears south of the Praesepe star cluster. (Cancer, evening sky.)
	N/A	Good opportunity to see Earthshine on the waxing crescent Moon. (Evening sky.)
18th	06:30	The waxing crescent Moon appears north of the bright star Regulus. (Leo, evening sky.)
19th	08:33	Neptune is stationary prior to beginning retrograde motion. (Morning sky.)
	16:37	Asteroid 4 Vesta is at opposition. (All night.)
	23:28	Venus appears 0.8° north of the Praesepe star cluster. (Cancer, evening sky.)
20th	10:52	First Quarter Moon. (Evening sky.)

Planet Locations – June 15th

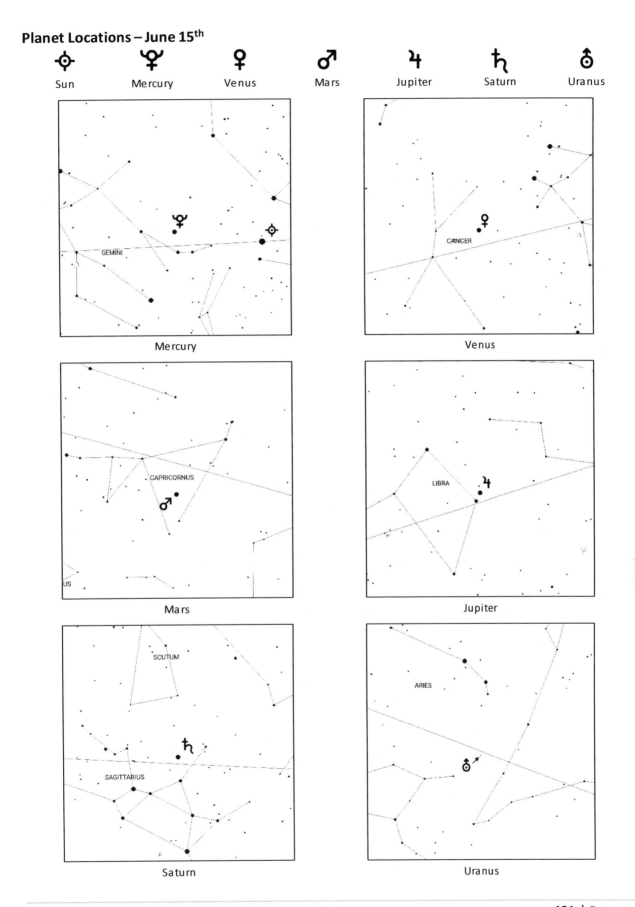

June 21st – 30th

The Moon

| | 21st | | 23rd | | 25th | | 27th | | 29th | |

Date	Con	R.A.	Dec	Mag	Diam	Ill.	Elong	Phase		Close To
21st	Vir	12h 56m	-1° 0'	-10.3	31'	61%	103° E	FQ		Spica
22nd	Vir	13h 44m	-5° 18'	-10.7	31'	71%	115° E	+G		Spica
23rd	Lib	14h 33m	-9° 33'	-11.1	30'	80%	127° E	+G		Jupiter
24th	Lib	15h 22m	-13° 18'	-11.4	30'	87%	138° E	+G		Jupiter
25th	Sco	16h 11m	-16° 24'	-11.7	30'	93%	150° E	+G		Antares
26th	Oph	17h 01m	-18° 42'	-12.1	29'	97%	161° E	FM		Antares
27th	Sgr	17h 52m	-20° 12'	-12.4	29'	100%	172° E	FM		Saturn
28th	Sgr	18h 43m	-20° 45'	-12.6	29'	100%	176° E	FM		Saturn
29th	Sgr	19h 33m	-20° 24'	-12.2	29'	99%	166° W	FM		
30th	Cap	20h 24m	-19° 6'	-11.9	29'	95%	155° W	-G		Mars

Mercury and Venus

Mercury
23rd

Mercury
27th

Venus
25th

Mercury

Date	Con	R.A.	Dec	Mag	Diam	Ill.	Elong	AM/PM		Close To
21st	Gem	07h 15m	24° 18'	-0.7	6''	78%	17° E	PM		
23rd	Gem	07h 31m	23° 45'	-0.5	6''	75%	19° E	PM		
25th	Gem	07h 45m	23° 6'	-0.4	6''	71%	20° E	PM		
27th	Gem	07h 59m	22° 21'	-0.2	6''	67%	22° E	PM		
29th	Cnc	08h 12m	21° 33'	-0.1	6''	64%	23° E	PM		

Venus

Date	Con	R.A.	Dec	Mag	Diam	Ill	Elong	AM/PM	Close To
21st	Cnc	08h 47m	19° 57'	-4.0	15''	74%	39° E	PM	Praesepe
23rd	Cnc	08h 57m	19° 18'	-4.0	15''	73%	39° E	PM	Praesepe
25th	Cnc	09h 06m	18° 36'	-4.0	15''	72%	40° E	PM	
27th	Cnc	09h 15m	17° 53'	-4.0	15''	71%	40° E	PM	
29th	Leo	09h 24m	17° 8'	-4.1	16''	71%	40° E	PM	

Mars and the Outer Planets

Mars
25th

Jupiter
25th

Saturn
25th

Mars

Date	Con	R.A.	Dec	Mag	Diam	Ill	Elong	AM/PM	Close To
20th	Cap	20h 50m	-22° 3'	-1.8	19''	95%	140° W	AM	
25th	Cap	20h 52m	-22° 21'	-2.0	20''	96%	144° W	AM	
30th	Cap	20h 52m	-22° 45'	-2.1	21''	97%	149° W	AM	Moon

The Outer Planets

Planet	Date	Con	R.A.	Dec	Mag	Diam	Elong	AM/PM	Close To
Jupiter	25th	Lib	14h 46m	-14° 51'	-2.3	42''	130° E	PM	
Saturn	25th	Sgr	18h 26m	-22° 27'	0.0	18''	178° W	AM	
Uranus	25th	Ari	01h 59m	11° 36'	5.8	3''	62° W	AM	
Neptune	25th	Aqr	23h 12m	-6° 12'	7.9	2''	107° W	AM	

Events

Date	Time (UT)	Event
21st	10:08	Estival solstice. Summer begins in the northern hemisphere, winter begins in the south.
22nd	03:56	The waxing gibbous Moon appears north of the bright star Spica. (Virgo, evening sky.)
23rd	17:20	The waxing gibbous Moons appears north of Jupiter. (Evening sky.)
25th	19:42	The waxing gibbous Moon appears north of the bright star Antares. (Scorpius, evening sky.)
27th	12:14	Saturn is at opposition. (All night.)
28th	04:54	Full Moon. (All night.)
	05:30	The full Moon appears north of Saturn. (All night.)
	13:45	Mars is stationary prior to beginning retrograde motion. (Morning sky.)
29th	10:03	The just-past full Moon appears north of the dwarf planet Pluto. (All night.)

Planet Locations – June 25th

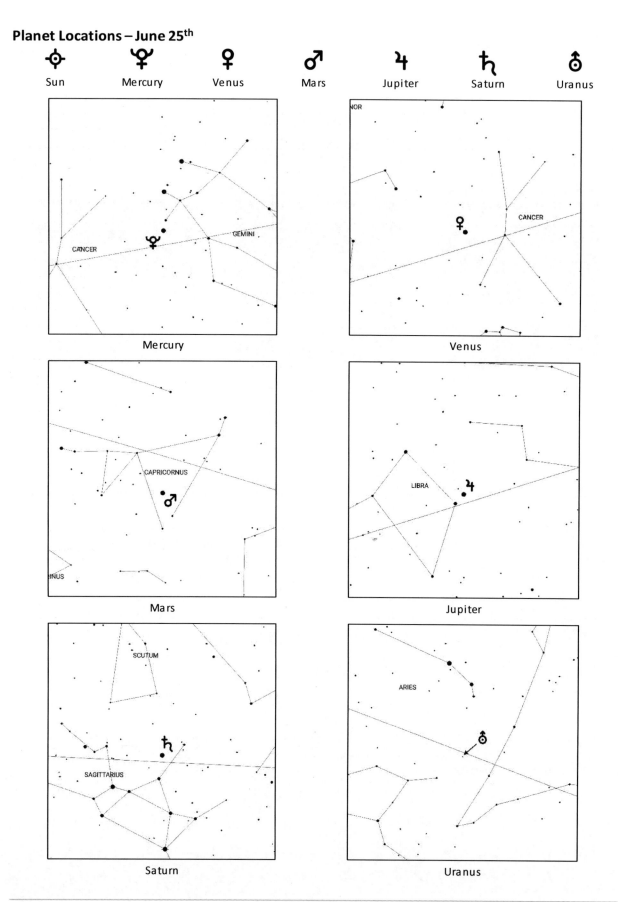

The Moon

1ˢᵗ   3ʳᵈ   5ᵗʰ   7ᵗʰ   9ᵗʰ

Date	Con	R.A.	Dec	Mag	Diam	Ill.	Elong	Phase	Close To
1st	Cap	21h 13m	-17° 3'	-11.6	29'	91%	144° W	-G	Mars
2nd	Cap	22h 01m	-14° 12'	-11.3	29'	84%	133° W	-G	
3rd	Aqr	22h 48m	-10° 48'	-10.9	30'	77%	122° W	-G	Neptune
4th	Aqr	23h 34m	-6° 51'	-10.6	30'	68%	111° W	-G	Neptune
5th	Psc	00h 21m	-2° 36'	-10.2	30'	59%	100° W	LQ	
6th	Cet	01h 08m	1° 51'	-9.7	31'	48%	88° W	LQ	
7th	Psc	01h 57m	6° 24'	-9.2	31'	38%	76° W	LQ	Uranus
8th	Ari	02h 49m	10° 45'	-8.6	32'	28%	63° W	-Cr	
9th	Tau	03h 44m	14° 42'	-8.0	33'	18%	50° W	-Cr	Pleiades, Hyades
10th	Tau	04h 42m	17° 53'	-7.2	33'	10%	37° W	NM	Hyades, Aldebaran

Mercury and Venus

Mercury 3ʳᵈ   Mercury 7ᵗʰ   Venus 5ᵗʰ

Mercury

Date	Con	R.A.	Dec	Mag	Diam	Ill.	Elong	AM/PM	Close To
1st	Cnc	08h 24m	20° 38'	0.0	7''	60%	24° E	PM	Praesepe
3rd	Cnc	08h 35m	19° 45'	0.1	7''	57%	25° E	PM	Praesepe
5th	Cnc	08h 46m	18° 48'	0.2	7''	53%	25° E	PM	Praesepe
7th	Cnc	08h 55m	17° 51'	0.3	7''	50%	26° E	PM	Praesepe
9th	Cnc	09h 04m	16° 51'	0.4	8''	47%	26° E	PM	

Venus

Date	Con	R.A.	Dec	Mag	Diam	Ill	Elong	AM/PM	Close To
1st	Leo	09h 33m	16° 23'	-4.1	16"	70%	41° E	PM	Regulus
3rd	Leo	09h 42m	15° 36'	-4.1	16"	69%	41° E	PM	Regulus
5th	Leo	09h 51m	14° 48'	-4.1	16"	68%	41° E	PM	Regulus
7th	Leo	09h 59m	13° 57'	-4.1	17"	68%	42° E	PM	Regulus
9th	Leo	10h 08m	13° 5'	-4.1	17"	67%	42° E	PM	Regulus

Mars and the Outer Planets

Mars
5th

Jupiter
5th

Saturn
5th

Mars

Date	Con	R.A.	Dec	Mag	Diam	Ill	Elong	AM/PM	Close To
1st	Cap	20h 52m	-22° 51'	-2.2	21"	97%	150° W	AM	Moon
5th	Cap	20h 51m	-23° 12'	-2.3	22"	98%	154° W	AM	
10th	Cap	20h 48m	-23° 42'	-2.4	23"	98%	159° W	AM	

The Outer Planets

Planet	Date	Con	R.A.	Dec	Mag	Diam	Elong	AM/PM	Close To
Jupiter	5th	Lib	14h 45m	-14° 51'	-2.3	41"	120° E	PM	
Saturn	5th	Sgr	18h 23m	-22° 30'	0.1	18"	172° E	PM	
Uranus	5th	Ari	02h 01m	11° 42'	5.8	4"	71° W	AM	
Neptune	5th	Aqr	23h 12m	-6° 15'	7.9	2"	117° W	AM	

Events

Date	Time (UT)	Event
1st	01:23	The waning gibbous Moon appears north of Mars. (Morning sky.)
3rd	22:27	The waning gibbous Moon appears south of Neptune. (Morning sky.)
4th	09:25	Mercury appears 0.3° south of the Praesepe star cluster. (Cancer, evening sky.)
6th	07:51	Last Quarter Moon. (Morning sky.)
7th	15:13	The just-past last quarter Moon appears south of Uranus. (Morning sky.)
9th	N/A	Good opportunity to see Earthshine on the waning crescent Moon. (Morning sky.)
	14:31	Venus appears 1.1° north of the bright star Regulus. (Leo, evening sky.)
	15:06	The waning crescent Moon appears south of the Pleiades star cluster. (Morning sky.)
10th	09:33	The waning crescent Moon appears north of the bright star Aldebaran. (Morning sky.)

Planet Locations – July 5th

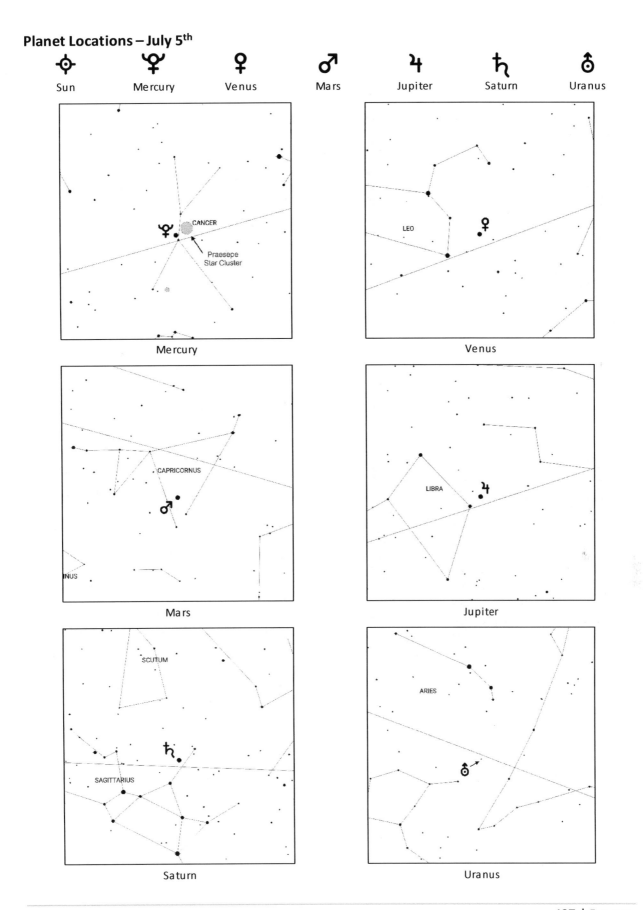

| Sun | Mercury | Venus | Mars | Jupiter | Saturn | Uranus |

Mercury

Venus

Mars

Jupiter

Saturn

Uranus

July 11th – 20th

The Moon

11th

13th

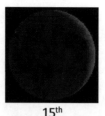
15th

17th

19th

Date	Con	R.A.	Dec	Mag	Diam	Ill.	Elong	Phase	Close To
11th	Tau	05h 44m	20° 0'	-6.2	34'	4%	23° W	NM	
12th	Gem	06h 48m	20° 45'	-5.1	34'	1%	9° W	NM	
13th	Gem	07h 53m	20° 0'	-4.8	34'	0%	6° E	NM	Praesepe
14th	Cnc	08h 57m	17° 48'	-6.0	34'	3%	20° E	NM	Mercury, Praesepe
15th	Leo	09h 58m	14° 24'	-6.9	34'	8%	34° E	NM	Mercury, Venus, Regulus
16th	Leo	10h 56m	10° 9'	-7.8	33'	16%	47° E	+Cr	Venus, Regulus
17th	Vir	11h 50m	5° 30'	-8.5	32'	26%	61° E	+Cr	
18th	Vir	12h 42m	0° 39'	-9.1	32'	36%	74° E	+Cr	Spica
19th	Vir	13h 32m	-4° 3'	-9.7	31'	47%	86° E	FQ	Spica
20th	Vir	14h 21m	-8° 27'	-10.1	31'	57%	98° E	FQ	Jupiter

Mercury and Venus

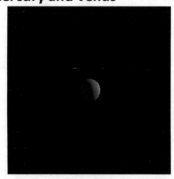
Mercury
13th

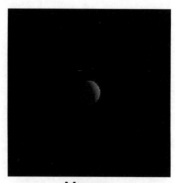
Mercury
17th

Venus
15th

Mercury

Date	Con	R.A.	Dec	Mag	Diam	Ill.	Elong	AM/PM	Close To
11th	Cnc	09h 12m	15° 54'	0.5	8''	44%	26° E	PM	
13th	Cnc	09h 19m	14° 57'	0.6	8''	41%	26° E	PM	
15th	Leo	09h 25m	14° 3'	0.7	8''	37%	26° E	PM	Moon
17th	Leo	09h 30m	13° 12'	0.8	9''	34%	26° E	PM	Regulus
19th	Leo	09h 34m	12° 24'	1.0	9''	30%	25° E	PM	Regulus

Venus

Date	Con	R.A.	Dec	Mag	Diam	Ill	Elong	AM/PM	Close To
11th	Leo	10h 16m	12° 12'	-4.1	17''	66%	42° E	PM	Regulus
13th	Leo	10h 25m	11° 18'	-4.1	17''	65%	43° E	PM	Regulus
15th	Leo	10h 33m	10° 24'	-4.1	18''	64%	43° E	PM	Moon, Regulus
17th	Leo	10h 41m	9° 30'	-4.1	18''	63%	43° E	PM	Regulus
19th	Leo	10h 49m	8° 33'	-4.1	18''	63%	44° E	PM	

Mars and the Outer Planets

Mars
15th

Jupiter
15th

Saturn
15th

Mars

Date	Con	R.A.	Dec	Mag	Diam	Ill	Elong	AM/PM	Close To
10th	Cap	20h 48m	-23° 42'	-2.4	23''	98%	159° W	AM	
15th	Cap	20h 45m	-24° 15'	-2.6	23''	99%	165° W	AM	
20th	Cap	20h 40m	-24° 48'	-2.7	24''	100%	170° W	AM	

The Outer Planets

Planet	Date	Con	R.A.	Dec	Mag	Diam	Elong	AM/PM	Close To
Jupiter	15th	Lib	14h 45m	-14° 51'	-2.2	40''	110° E	PM	
Saturn	15th	Sgr	18h 20m	-22° 33'	0.1	18''	162° E	PM	
Uranus	15th	Ari	02h 01m	11° 48'	5.8	4''	81° W	AM	
Neptune	15th	Aqr	23h 11m	-6° 18'	7.8	2''	127° W	AM	

Events

Date	Time (UT)	Event
11th	03:25	Jupiter is stationary prior to resuming prograde motion. (Evening sky.)
12th	04:33	Dwarf planet Pluto is at opposition. (All night.)
	05:19	Mercury is at greatest eastern elongation. (Evening sky.)
13th	02:49	New Moon. (Not visible.)
	03:02	Partial solar eclipse. (South Australia, Indian Ocean and Pacific Ocean.)
14th	23:14	The just-past new Moon appears north of Mercury. (Evening sky.)
15th	17:27	The waxing crescent Moon appears north of the bright star Regulus. (Evening sky.)
16th	03:23	The waxing crescent Moon appears north of Venus. (Evening sky.)
	N/A	Good opportunity to see Earthshine on the waxing crescent Moon. (Evening sky.)
19th	07:44	The almost first quarter Moon appears north of the bright star Spica. (Evening sky.)
	19:53	First Quarter Moon. (Evening sky.)

Planet Locations – July 15th

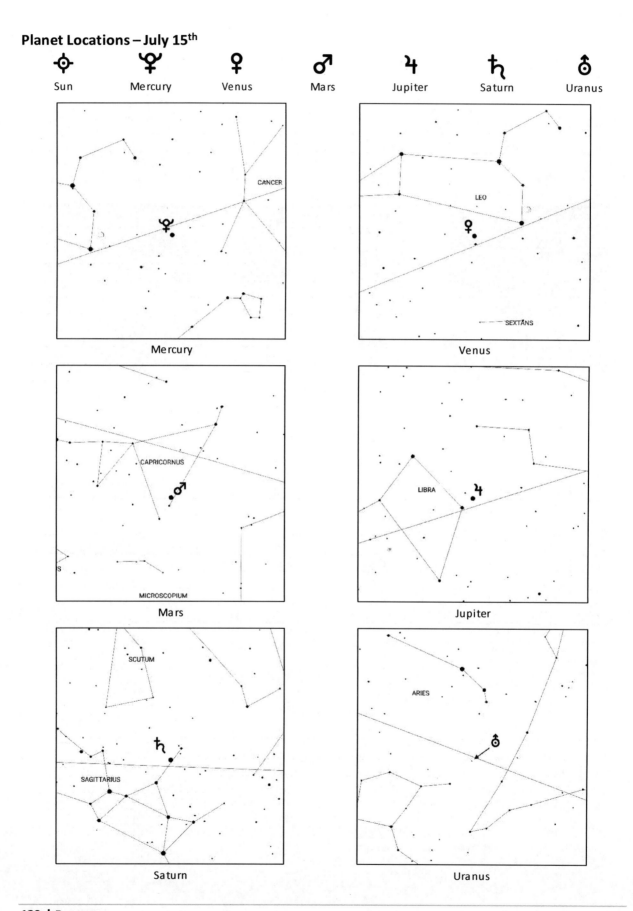

| Sun | Mercury | Venus | Mars | Jupiter | Saturn | Uranus |

The Moon

| 21st | 23rd | 25th | 27th | 29th | 31st |

Date	Con	R.A.	Dec	Mag	Diam	Ill.	Elong	Phase	Close To
21st	Lib	15h 10m	-12° 21'	-10.5	30'	67%	110° E	+G	Jupiter
22nd	Sco	15h 59m	-15° 39'	-10.9	30'	76%	121° E	+G	Antares
23rd	Oph	16h 49m	-18° 9'	-11.2	30'	84%	132° E	+G	Antares
24th	Oph	17h 39m	-19° 54'	-11.6	29'	90%	143° E	+G	Saturn
25th	Sgr	18h 30m	-20° 42'	-11.9	29'	95%	154° E	FM	Saturn
26th	Sgr	19h 21m	-20° 33'	-12.2	29'	98%	165° E	FM	
27th	Cap	20h 11m	-19° 33'	-12.6	29'	100%	176° E	FM	Mars
28th	Cap	21h 00m	-17° 39'	-12.5	29'	100%	173° W	FM	Mars
29th	Cap	21h 49m	-15° 0'	-12.1	29'	98%	162° W	FM	
30th	Aqr	22h 36m	-11° 42'	-11.8	29'	94%	151° W	-G	
31st	Aqr	23h 23m	-7° 54'	-11.5	30'	88%	140° W	-G	Neptune

Mercury and Venus

  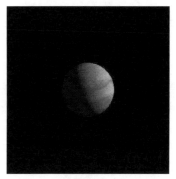

Mercury 23rd   Mercury 27th   Venus 25th

Mercury

Date	Con	R.A.	Dec	Mag	Diam	Ill.	Elong	AM/PM	Close To
21st	Leo	09h 36m	11° 42'	1.1	9''	27%	24° E	PM	Regulus
23rd	Leo	09h 38m	11° 3'	1.3	10''	23%	23° E	PM	Regulus
25th	Leo	09h 39m	10° 33'	1.6	10''	20%	21° E	PM	Regulus
27th	Leo	09h 37m	10° 9'	1.9	10''	16%	19° E	PM	Regulus
29th	Leo	09h 36m	9° 54'	2.2	11''	12%	17° E	PM	Regulus
31st	Leo	09h 33m	9° 51'	2.7	11''	9%	15° E	NV	Regulus

Venus

Date	Con	R.A.	Dec	Mag	Diam	Ill	Elong	AM/PM	Close To
21st	Leo	10h 57m	7° 36'	-4.1	19''	62%	44° E	PM	
23rd	Leo	11h 05m	6° 39'	-4.2	19''	61%	44° E	PM	
25th	Leo	11h 13m	5° 42'	-4.2	19''	60%	44° E	PM	
27th	Leo	11h 20m	4° 42'	-4.2	20''	59%	45° E	PM	
29th	Leo	11h 28m	3° 45'	-4.2	20''	58%	45° E	PM	
31st	Leo	11h 36m	2° 45'	-4.2	20''	57%	45° E	PM	

Mars and the Outer Planets

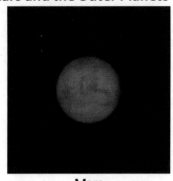

Mars
25th

Jupiter
25th

Saturn
25th

Mars

Date	Con	R.A.	Dec	Mag	Diam	Ill	Elong	AM/PM	Close To
20th	Cap	20h 40m	-24° 48'	-2.7	24''	100%	170° W	AM	
25th	Cap	20h 35m	-25° 21'	-2.8	24''	100%	173° W	AM	
30th	Cap	20h 29m	-25° 48'	-2.8	24''	100%	172° E	PM	

The Outer Planets

Planet	Date	Con	R.A.	Dec	Mag	Diam	Elong	AM/PM	Close To
Jupiter	25th	Lib	14h 46m	-15° 0'	-2.1	39''	101° E	PM	
Saturn	25th	Sgr	18h 17m	-22° 36'	0.2	18''	151° E	PM	Moon
Uranus	25th	Ari	02h 02m	11° 51'	5.8	4''	90° W	AM	
Neptune	25th	Aqr	23h 11m	-6° 21'	7.8	2''	136° W	AM	

Events

Date	Time (UT)	Event
21st	01:44	The waxing gibbous Moon appears north of Jupiter. (Evening sky.)
23rd	04:13	The waxing gibbous Moon appears north of the bright star Antares. (Scorpius, evening sky.)
25th	07:09	The waxing gibbous Moon appears north of Saturn. (Evening sky.)
	07:18	Mercury is stationary prior to beginning retrograde motion. (Evening sky.)
26th	12:47	The nearly full Moon appears north of dwarf planet Pluto. (All night.)
27th	20:21	Full Moon. (All night.)
	20:22	Total lunar eclipse. (Africa, Asia, Australia, Europe, Indian Ocean, Pacific Ocean and South America)
	20:40	The full Moon appears north of Mars. (All night.)
28th	13:07	Mars is at opposition. (All night)
29th	N/A	The Delta Aquariid meteors are at their peak. (ZHR: 16. All night but best in the morning sky.)
31st	07:21	The waning gibbous Moon appears south of Neptune. (Morning sky.)

Planet Locations – July 25th

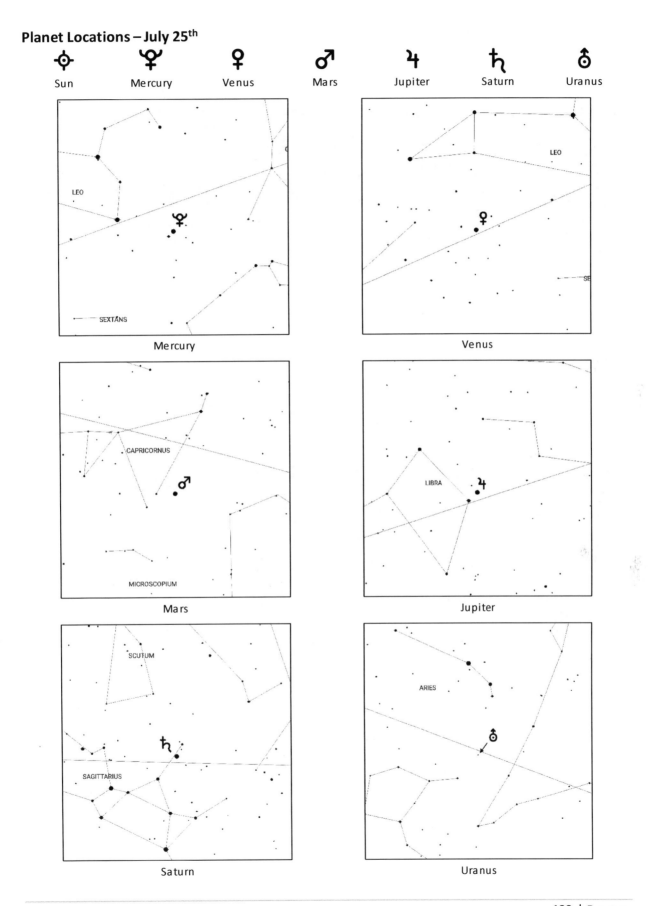

August 1st to 10th

The Moon

| | 1st | | 3rd | | 5th | | 7th | | 9th |

Date	Con	R.A.	Dec	Mag	Diam	Ill.	Elong	Phase	Close To
1st	Psc	00h 09m	-3° 45'	-11.1	30'	81%	129° W	-G	
2nd	Cet	00h 56m	0° 39'	-10.8	30'	73%	117° W	-G	
3rd	Psc	01h 43m	5° 5'	-10.4	31'	63%	105° W	LQ	Uranus
4th	Cet	02h 33m	9° 24'	-9.9	31'	53%	93° W	LQ	
5th	Ari	03h 25m	13° 24'	-9.4	32'	42%	81° W	LQ	Pleiades
6th	Tau	04h 20m	16° 48'	-8.9	32'	31%	68° W	-Cr	Pleiades, Hyades, Aldebaran
7th	Tau	05h 19m	19° 18'	-8.2	33'	21%	55° W	-Cr	Aldebaran
8th	Ori	06h 21m	20° 36'	-7.4	33'	12%	41° W	NM	
9th	Gem	07h 25m	20° 30'	-6.5	34'	6%	27° W	NM	
10th	Cnc	08h 29m	18° 57'	-5.4	34'	1%	13° W	NM	Mercury, Praesepe

Mercury and Venus

Mercury	Mercury	Venus
3rd	7th	5th

Mercury

Date	Con	R.A.	Dec	Mag	Diam	Ill.	Elong	AM/PM	Close To
1st	Leo	09h 31m	9° 51'	2.9	11''	8%	13° E	NV	Regulus
3rd	Leo	09h 26m	10° 3'	3.4	11''	5%	11° E	NV	
5th	Cnc	09h 20m	10° 21'	4.0	11''	3%	8° E	NV	
7th	Cnc	09h 15m	10° 51'	4.6	11''	1%	6° E	NV	
9th	Cnc	09h 09m	11° 27'	4.7	11''	1%	5° W	NV	

Venus

Date	Con	R.A.	Dec	Mag	Diam	Ill	Elong	AM/PM	Close To
1st	Vir	11h 39m	2° 15'	-4.2	21''	57%	45° E	PM	
3rd	Vir	11h 47m	1° 15'	-4.2	21''	56%	45° E	PM	
5th	Vir	11h 54m	0° 15'	-4.2	21''	55%	46° E	PM	
7th	Vir	12h 01m	-1° 0'	-4.3	22''	54%	46° E	PM	
9th	Vir	12h 08m	-1° 39'	-4.3	22''	53%	46° E	PM	

Mars and the Outer Planets

Mars
5th

Jupiter
5th

Saturn
5th

Mars

Date	Con	R.A.	Dec	Mag	Diam	Ill	Elong	AM/PM	Close To
1st	Cap	20h 27m	-25° 57'	-2.8	24''	100%	171° E	PM	
5th	Cap	20h 22m	-26° 12'	-2.7	24''	99%	167° E	PM	
10th	Cap	20h 17m	-26° 24'	-2.6	24''	99%	162° E	PM	

The Outer Planets

Planet	Date	Con	R.A.	Dec	Mag	Diam	Elong	AM/PM	Close To
Jupiter	5th	Lib	14h 49m	-15° 12'	-2.1	37''	91° E	PM	
Saturn	5th	Sgr	18h 14m	-22° 36'	0.2	18''	140° E	PM	
Uranus	5th	Ari	02h 02m	11° 51'	5.8	4''	101° W	AM	
Neptune	5th	Aqr	23h 10m	-6° 27'	7.8	2''	147° W	AM	

Events

Date	Time (UT)	Event
3rd	20:08	The nearly last quarter Moon appears south of Uranus. (Morning sky.)
4th	18:19	Last Quarter Moon. (Evening sky.)
5th	20:56	The just-past last quarter Moon appears to the south of the Pleiades star cluster. (Taurus, morning sky.)
6th	18:53	The waning crescent Moon appears to the north of the bright star Aldebaran. (Taurus, evening sky.)
7th	16:54	Uranus is stationary prior to beginning retrograde motion. (Morning sky.)
8th	N/A	Good opportunity to see Earthshine on the waning crescent Moon. (Morning sky.)
9th	01:59	Mercury is at inferior conjunction with the Sun. (Not visible.)

Planet Locations – August 5ᵗʰ

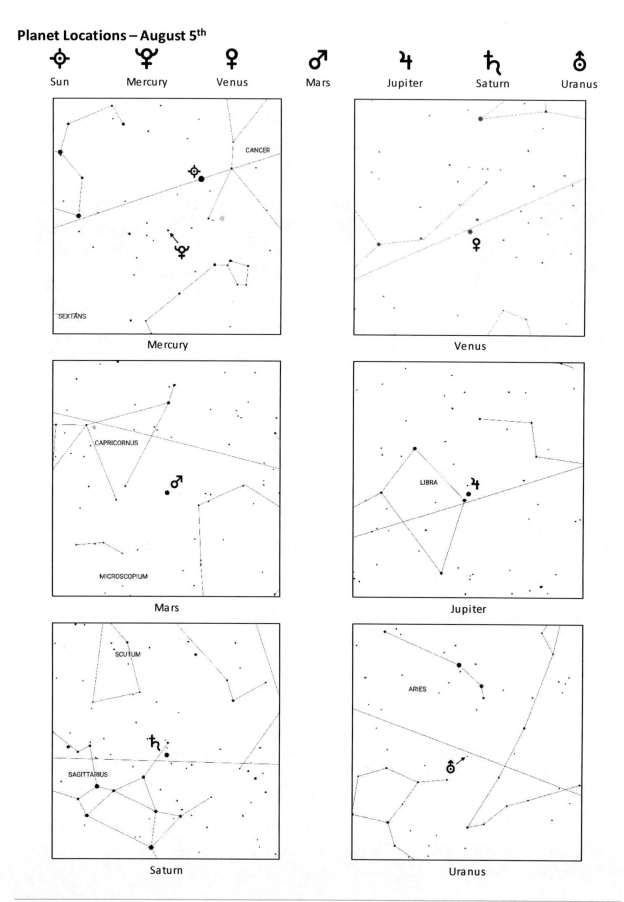

The Moon

11th 13th 15th 17th 19th

Date	Con	R.A.	Dec	Mag	Diam	Ill.	Elong	Phase	Close To
11th	Leo	09h 31m	16° 3'	-4.5	34'	0%	2° E	NM	Mercury, Regulus
12th	Leo	10h 31m	12° 3'	-5.6	34'	2%	16° E	NM	Regulus
13th	Leo	11h 28m	7° 24'	-6.7	33'	6%	29° E	NM	
14th	Vir	12h 22m	2° 27'	-7.5	33'	13%	43° E	+Cr	Venus
15th	Vir	13h 15m	-2° 30'	-8.3	32'	22%	56° E	+Cr	Venus, Spica
16th	Vir	14h 06m	-7° 6'	-8.9	32'	32%	68° E	+Cr	Jupiter, Spica
17th	Lib	14h 56m	-11° 18'	-9.4	31'	42%	80° E	FQ	Jupiter
18th	Lib	15h 46m	-14° 51'	-9.9	30'	52%	92° E	FQ	Antares
19th	Oph	16h 36m	-17° 36'	-10.3	30'	62%	103° E	FQ	Antares
20th	Oph	17h 26m	-19° 33'	-10.7	30'	71%	115° E	+G	Saturn

Mercury and Venus

  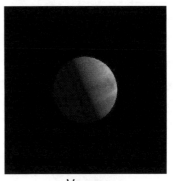

Mercury 13th Mercury 17th Venus 15th

Mercury

Date	Con	R.A.	Dec	Mag	Diam	Ill.	Elong	AM/PM	Close To
11th	Cnc	09h 03m	12° 5'	4.3	11''	2%	6° W	NV	Moon
13th	Cnc	08h 58m	12° 51'	3.6	10''	4%	9° W	NV	Praesepe
15th	Cnc	08h 55m	13° 33'	2.9	10''	7%	11° W	NV	Praesepe
17th	Cnc	08h 53m	14° 15'	2.2	10''	11%	13° W	NV	Praesepe
19th	Cnc	08h 53m	14° 48'	1.6	9''	16%	15° W	NV	Praesepe

Venus

Date	Con	R.A.	Dec	Mag	Diam	Ill	Elong	AM/PM		Close To
11th	Vir	12h 15m	-2° 39'	-4.3	23''	52%	46° E	PM		
13th	Vir	12h 22m	-3° 36'	-4.3	23''	51%	46° E	PM		
15th	Vir	12h 29m	-4° 36'	-4.3	24''	50%	46° E	PM		Moon
17th	Vir	12h 36m	-5° 33'	-4.3	24''	49%	46° E	PM		
19th	Vir	12h 43m	-6° 30'	-4.3	25''	48%	46° E	PM		

Mars and the Outer Planets

Mars
15th

Jupiter
15th

Saturn
15th

Mars

Date	Con	R.A.	Dec	Mag	Diam	Ill	Elong	AM/PM		Close To
10th	Cap	20h 17m	-26° 24'	-2.6	24''	99%	162° E	PM		
15th	Cap	20h 13m	-26° 30'	-2.5	23''	98%	156° E	PM		
20th	Cap	20h 10m	-26° 27'	-2.4	23''	97%	151° E	PM		

The Outer Planets

Planet	Date	Con	R.A.	Dec	Mag	Diam	Elong	AM/PM		Close To
Jupiter	15th	Lib	14h 52m	-15° 30'	-2.0	36''	83° E	PM		
Saturn	15th	Sgr	18h 13m	-22° 39'	0.3	18''	130° E	PM		
Uranus	15th	Ari	02h 02m	11° 51'	5.8	4''	110° W	AM		
Neptune	15th	Aqr	23h 09m	-6° 33'	7.8	2''	157° W	AM		

Events

Date	Time (UT)	Event
11th	09:48	Partial solar eclipse. (North and west Asia, north and east Europe, north-east North America.)
	09:58	New Moon (Not visible.)
12th	N/A	The Perseid meteor shower reaches its peak. (ZHR: 100. All night but best seen in the morning sky.)
14th	12:27	The waxing crescent Moon appears to the north of Venus. (Evening sky.)
	N/A	Good opportunity to see Earthshine on the waxing crescent Moon. (Evening sky.)
15th	17:47	The waxing crescent Moon appears to the north of the bright star Spica. (Virgo, evening sky.)
17th	09:03	The nearly first quarter Moon appears to the north of Jupiter. (Evening sky.)
	17:26	Venus is at greatest eastern elongation. (Evening sky.)
18th	07:49	First Quarter Moon. (Evening sky.)
	12:22	Mercury is stationary prior to resuming prograde motion. (Not visible.)
19th	08:14	The just-past first quarter Moon appears to the north of the bright star Antares. (Scorpius, evening sky.)

Planet Locations – August 15th

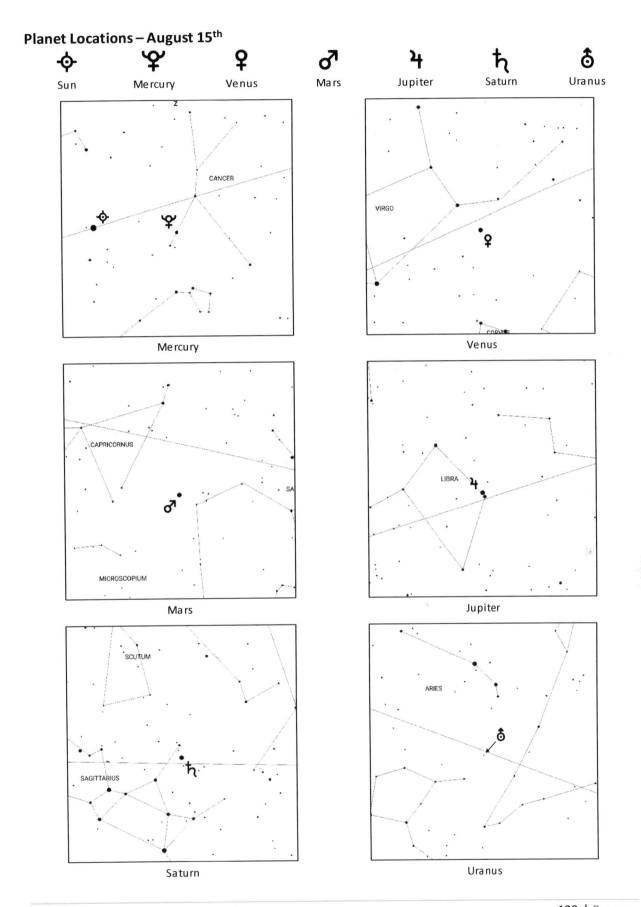

⊕	☿	♀	♂	♃	♄	♅
Sun	Mercury	Venus	Mars	Jupiter	Saturn	Uranus

Mercury

Venus

Mars

Jupiter

Saturn

Uranus

August 21st – 31st

The Moon

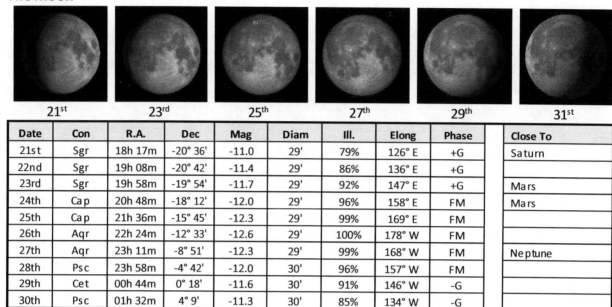

| 21st | 23rd | 25th | 27th | 29th | 31st |

Date	Con	R.A.	Dec	Mag	Diam	Ill.	Elong	Phase		Close To
21st	Sgr	18h 17m	-20° 36'	-11.0	29'	79%	126° E	+G		Saturn
22nd	Sgr	19h 08m	-20° 42'	-11.4	29'	86%	136° E	+G		
23rd	Sgr	19h 58m	-19° 54'	-11.7	29'	92%	147° E	+G		Mars
24th	Cap	20h 48m	-18° 12'	-12.0	29'	96%	158° E	FM		Mars
25th	Cap	21h 36m	-15° 45'	-12.3	29'	99%	169° E	FM		
26th	Aqr	22h 24m	-12° 33'	-12.6	29'	100%	178° W	FM		
27th	Aqr	23h 11m	-8° 51'	-12.3	29'	99%	168° W	FM		Neptune
28th	Psc	23h 58m	-4° 42'	-12.0	30'	96%	157° W	FM		
29th	Cet	00h 44m	0° 18'	-11.6	30'	91%	146° W	-G		
30th	Psc	01h 32m	4° 9'	-11.3	30'	85%	134° W	-G		
31st	Cet	02h 20m	8° 30'	-10.9	31'	77%	122° W	-G		Uranus

Mercury and Venus

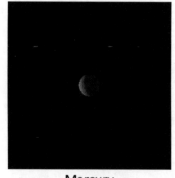

Mercury
23rd

Mercury
27th

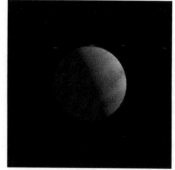

Venus
25th

Mercury

Date	Con	R.A.	Dec	Mag	Diam	Ill.	Elong	AM/PM		Close To
21st	Cnc	08h 55m	15° 18'	1.0	9''	23%	17° W	AM		Praesepe
23rd	Cnc	08h 59m	15° 39'	0.5	8''	30%	18° W	AM		Praesepe
25th	Cnc	09h 05m	15° 48'	0.1	8''	38%	18° W	AM		
27th	Cnc	09h 13m	15° 48'	-0.2	7''	46%	18° W	AM		
29th	Leo	09h 22m	15° 33'	-0.5	7''	54%	18° W	AM		
31st	Leo	09h 34m	15° 5'	-0.7	6''	63%	17° W	AM		Regulus

Venus

Date	Con	R.A.	Dec	Mag	Diam	Ill	Elong	AM/PM	Close To
21st	Vir	12h 49m	-7° 24'	-4.4	26''	47%	46° E	PM	Spica
23rd	Vir	12h 56m	-8° 21'	-4.4	26''	46%	46° E	PM	Spica
25th	Vir	13h 02m	-9° 15'	-4.4	27''	44%	46° E	PM	Spica
27th	Vir	13h 09m	-10° 9'	-4.4	28''	43%	46° E	PM	Spica
29th	Vir	13h 15m	-11° 3'	-4.4	28''	42%	45° E	PM	Spica
31st	Vir	13h 21m	-11° 54'	-4.4	29''	41%	45° E	PM	Spica

Mars and the Outer Planets

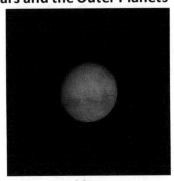

Mars
25th

Jupiter
25th

Saturn
25th

Mars

Date	Con	R.A.	Dec	Mag	Diam	Ill	Elong	AM/PM	Close To
20th	Cap	20h 10m	-26° 27'	-2.4	23''	97%	151° E	PM	
25th	Cap	20h 09m	-26° 18'	-2.3	22''	96%	146° E	PM	
30th	Cap	20h 08m	-26° 3'	-2.1	21''	95%	141° E	PM	

The Outer Planets

Planet	Date	Con	R.A.	Dec	Mag	Diam	Elong	AM/PM	Close To
Jupiter	25th	Lib	14h 56m	-15° 51'	-2.0	35''	74° E	PM	
Saturn	25th	Sgr	18h 12m	-22° 39'	0.4	17''	120° E	PM	
Uranus	25th	Ari	02h 02m	11° 48'	5.7	4''	120° W	AM	
Neptune	25th	Aqr	23h 08m	-6° 39'	7.8	2''	167° W	AM	

Events

Date	Time (UT)	Event
21st	09:12	The waxing gibbous Moon appears to the north of Saturn. (Evening sky.)
22nd	17:00	The waxing gibbous Moon appears to the north of dwarf planet Pluto. (Evening sky.)
23rd	15:27	The waxing gibbous Moon appears to the north of Mars. (Evening sky.)
26th	11:57	Full Moon. (All night.)
	20:27	Mercury is at greatest western elongation. (Morning sky.)
27th	11:02	The just-past full Moon appears to the south of Neptune. (All night.)
28th	10:16	Mars is stationary prior to resuming prograde motion. (Evening sky.)
31st	02:07	The waning gibbous Moon appears to the south of Uranus. (Morning sky.)

Planet Locations – August 25th

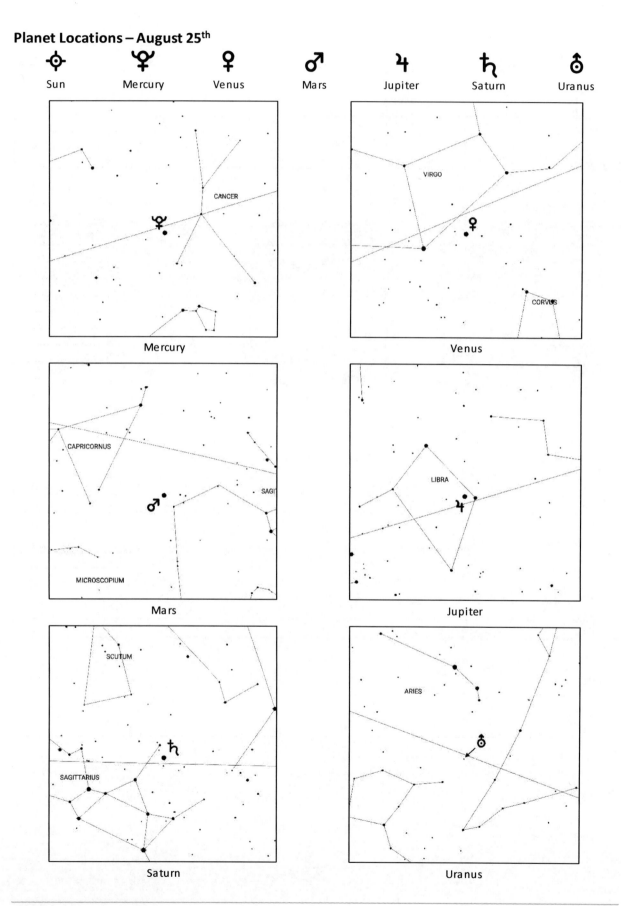

The Moon

1st   3rd   5th   7th   9th

Date	Con	R.A.	Dec	Mag	Diam	Ill.	Elong	Phase	Close To
1st	Ari	03h 11m	12° 33'	-10.5	31'	67%	110° W	-G	Pleiades
2nd	Tau	04h 04m	16° 3'	-10.1	32'	57%	98° W	LQ	Pleiades, Hyades, Aldebaran
3rd	Tau	05h 01m	18° 45'	-9.6	32'	46%	85° W	LQ	Hyades, Aldebaran
4th	Ori	06h 00m	20° 23'	-9.1	33'	35%	72° W	-Cr	
5th	Gem	07h 01m	20° 48'	-8.4	33'	24%	59° W	-Cr	
6th	Gem	08h 03m	19° 48'	-7.7	33'	15%	45° W	-Cr	Praesepe
7th	Cnc	09h 05m	17° 27'	-6.8	34'	7%	31° W	NM	Praesepe
8th	Leo	10h 05m	13° 54'	-5.8	34'	2%	18° W	NM	Mercury, Regulus
9th	Leo	11h 03m	9° 30'	-4.7	34'	0%	5° W	NM	Mercury
10th	Vir	11h 59m	4° 33'	-5.3	33'	1%	11° E	NM	

Mercury and Venus

Mercury   Mercury   Venus

3rd   7th   5th

Mercury

Date	Con	R.A.	Dec	Mag	Diam	Ill.	Elong	AM/PM	Close To
1st	Leo	09h 40m	14° 48'	-0.8	6''	67%	17° W	AM	Regulus
3rd	Leo	09h 53m	14° 5'	-1.0	6''	74%	16° W	AM	Regulus
5th	Leo	10h 06m	13° 9'	-1.1	6''	81%	14° W	NV	Regulus
7th	Leo	10h 20m	12° 3'	-1.2	6''	86%	12° W	NV	Regulus
9th	Leo	10h 35m	10° 48'	-1.3	5''	91%	11° W	NV	Moon, Regulus

Venus

Date	Con	R.A.	Dec	Mag	Diam	Ill	Elong	AM/PM	Close To
1st	Vir	13h 24m	-12° 18'	-4.4	29''	40%	45° E	PM	Spica
3rd	Vir	13h 29m	-13° 9'	-4.5	30''	39%	45° E	PM	Spica
5th	Vir	13h 35m	-13° 57'	-4.5	31''	37%	44° E	PM	Spica
7th	Vir	13h 40m	-14° 45'	-4.5	32''	36%	44° E	PM	Spica
9th	Vir	13h 46m	-15° 30'	-4.5	33''	35%	44° E	PM	Spica

Mars and the Outer Planets

Mars
5th

Jupiter
5th

Saturn
5th

Mars

Date	Con	R.A.	Dec	Mag	Diam	Ill	Elong	AM/PM	Close To
1st	Cap	20h 09m	-25° 57'	-2.1	21''	94%	139° E	PM	
5th	Cap	20h 10m	-25° 36'	-2.0	20''	93%	136° E	PM	
10th	Cap	20h 13m	-25° 9'	-1.9	19''	92%	132° E	PM	

The Outer Planets

Planet	Date	Con	R.A.	Dec	Mag	Diam	Elong	AM/PM	Close To
Jupiter	5th	Lib	15h 02m	-16° 21'	-1.9	34''	65° E	PM	
Saturn	5th	Sgr	18h 11m	-22° 42'	0.4	17''	110° E	PM	
Uranus	5th	Ari	02h 01m	11° 42'	5.7	4''	131° W	AM	
Neptune	5th	Aqr	23h 07m	-6° 48'	7.8	2''	178° W	AM	

Events

Date	Time (UT)	Event
2nd	00:56	Venus appears 1.4° south of the bright star Spica. (Virgo, evening sky.)
	03:43	The nearly last quarter Moon appears south of the Pleiades star cluster. (Morning sky.)
	23:52	The nearly last quarter Moon appears north of the bright star Aldebaran. (Morning sky.)
3rd	02:38	Last Quarter Moon. (Morning sky.)
6th	N/A	Good opportunity to see Earthshine on the waning crescent Moon. (Morning sky.)
	09:26	Saturn is stationary prior to resuming prograde motion. (Evening sky.)
8th	06:20	Neptune is at opposition. (All night.)
9th	18:02	New Moon. (Not visible.)

Planet Locations – September 5th

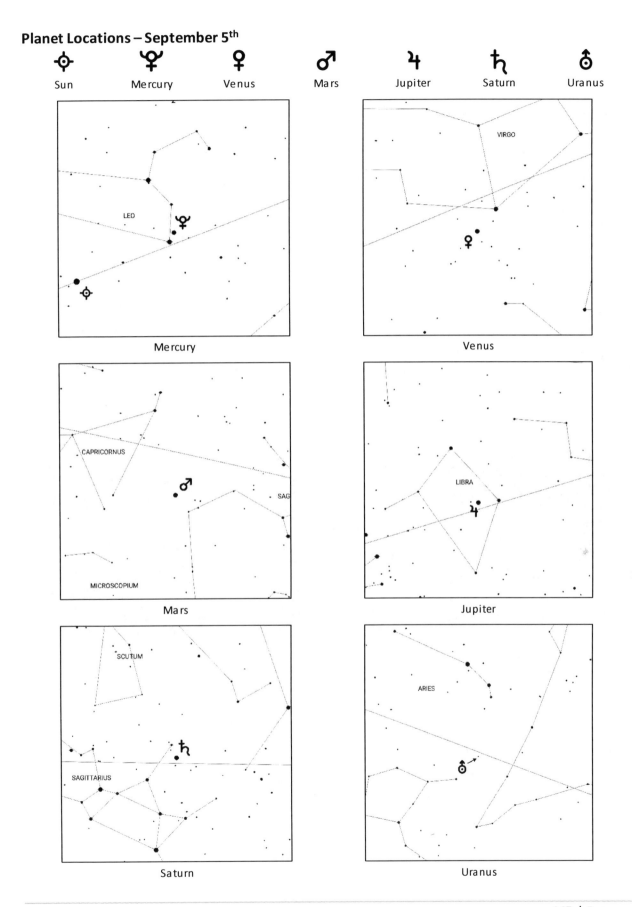

| Sun | Mercury | Venus | Mars | Jupiter | Saturn | Uranus |

September 11th – 20th

The Moon

| 11th | 13th | 15th | 17th | 19th |

Date	Con	R.A.	Dec	Mag	Diam	Ill.	Elong	Phase	Close To
11th	Vir	12h 53m	-1° 0'	-6.3	33'	4%	24° E	NM	Spica
12th	Vir	13h 45m	-5° 27'	-7.2	32'	10%	37° E	NM	Venus, Spica
13th	Lib	14h 37m	-9° 57'	-7.9	32'	18%	50° E	+Cr	Venus, Jupiter
14th	Lib	15h 28m	-13° 48'	-8.6	31'	26%	62° E	+Cr	Jupiter
15th	Sco	16h 19m	-16° 57'	-9.1	31'	36%	73° E	+Cr	Antares
16th	Oph	17h 11m	-19° 12'	-9.6	30'	46%	85° E	FQ	Antares
17th	Sgr	18h 02m	-20° 30'	-10.0	30'	55%	96° E	FQ	Saturn
18th	Sgr	18h 53m	-20° 54'	-10.4	30'	65%	107° E	FQ	Saturn
19th	Sgr	19h 44m	-20° 21'	-10.8	29'	73%	118° E	+G	Mars
20th	Cap	20h 33m	-18° 54'	-11.1	29'	81%	129° E	+G	Mars

Mercury and Venus

| Mercury 13th | Mercury 17th | Venus 15th |

Mercury

Date	Con	R.A.	Dec	Mag	Diam	Ill.	Elong	AM/PM	Close To
11th	Leo	10h 49m	9° 27'	-1.4	5''	94%	9° W	NV	
13th	Leo	11h 03m	8° 0'	-1.5	5''	97%	7° W	NV	
15th	Leo	11h 17m	6° 30'	-1.5	5''	98%	5° W	NV	
17th	Leo	11h 31m	5° 0'	-1.6	5''	99%	4° W	NV	
19th	Vir	11h 45m	3° 24'	-1.6	5''	100%	2° W	NV	

Venus

Date	Con	R.A.	Dec	Mag	Diam	Ill	Elong	AM/PM		Close To
11th	Vir	13h 51m	-16° 15'	-4.5	34''	33%	43° E	PM		Spica
13th	Vir	13h 56m	-16° 57'	-4.5	35''	32%	42° E	PM		Moon, Spica
15th	Vir	14h 00m	-17° 36'	-4.5	36''	30%	42° E	PM		Spica
17th	Vir	14h 04m	-18° 15'	-4.5	37''	29%	41° E	PM		Spica
19th	Vir	14h 08m	-18° 51'	-4.6	38''	27%	40° E	PM		

Mars and the Outer Planets

Mars
15th

Jupiter
15th

Saturn
15th

Mars

Date	Con	R.A.	Dec	Mag	Diam	Ill	Elong	AM/PM		Close To
10th	Cap	20h 13m	-25° 9'	-1.9	19''	92%	132° E	PM		
15th	Cap	20h 17m	-24° 36'	-1.7	18''	91%	128° E	PM		
20th	Cap	20h 23m	-24° 3'	-1.6	18''	90%	125° E	PM		Moon

The Outer Planets

Planet	Date	Con	R.A.	Dec	Mag	Diam	Elong	AM/PM		Close To
Jupiter	15th	Lib	15h 09m	-16° 48'	-1.9	34''	57° E	PM		
Saturn	15th	Sgr	18h 11m	-22° 45'	0.5	17''	100° E	PM		
Uranus	15th	Ari	02h 00m	11° 39'	5.7	4''	141° W	AM		
Neptune	15th	Aqr	23h 06m	-6° 51'	7.8	2''	172° E	PM		

Events

Date	Time (UT)	Event
12th	19:37	The waxing crescent Moon appears north of the bright star Spica. (Virgo, evening sky.)
13th	N/A	Good opportunity to see Earthshine on the waxing crescent Moon. (Evening sky.)
14th	02:44	The waxing crescent Moon appears north of Jupiter. (Evening sky.)
15th	16:36	The nearly first quarter Moon appears north of the bright star Antares. (Scorpius, evening sky.)
16th	23:16	First Quarter Moon. (Evening sky.)
17th	15:11	The just-past first quarter Moon appears north of Saturn. (Evening sky.)
19th	02:41	The waxing gibbous Moon appears north of dwarf planet Pluto. (Evening sky.)
20th	07:17	The waxing gibbous Moon appears north of Mars. (Evening sky.)

Planet Locations – September 15th

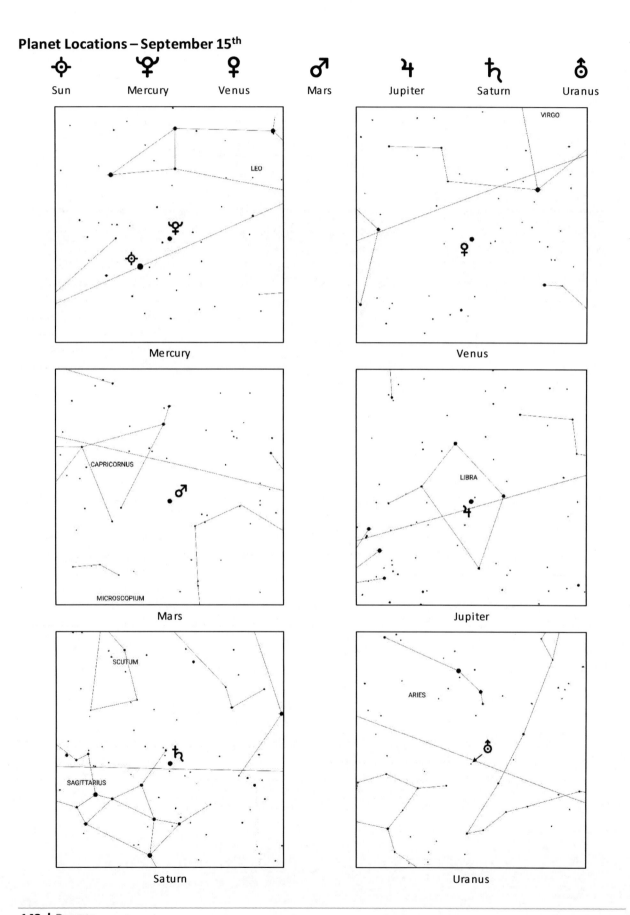

| Sun | Mercury | Venus | Mars | Jupiter | Saturn | Uranus |

Mercury

Venus

Mars

Jupiter

Saturn

Uranus

The Moon

| 21st | 23rd | 25th | 27th | 29th |

Date	Con	R.A.	Dec	Mag	Diam	Ill.	Elong	Phase	Close To
21st	Cap	21h 22m	-16° 36'	-11.4	29'	88%	139° E	+G	
22nd	Aqr	22h 11m	-13° 36'	-11.8	29'	93%	150° E	+G	
23rd	Aqr	22h 58m	-9° 57'	-12.1	30'	97%	161° E	FM	Neptune
24th	Aqr	23h 45m	-5° 51'	-12.4	30'	100%	172° E	FM	
25th	Cet	00h 32m	-1° 27'	-12.5	30'	100%	174° W	FM	
26th	Psc	01h 20m	3° 6'	-12.2	30'	98%	163° W	FM	
27th	Cet	02h 09m	7° 36'	-11.8	31'	94%	151° W	-G	Uranus
28th	Ari	02h 59m	11° 48'	-11.4	31'	88%	139° W	-G	Pleiades
29th	Tau	03h 52m	15° 30'	-11.1	31'	80%	127° W	-G	Pleiades, Hyades, Aldebaran
30th	Tau	04h 47m	18° 23'	-10.7	32'	71%	115° W	-G	Hyades, Aldebaran

Mercury and Venus

Mercury
23rd

Mercury
27th

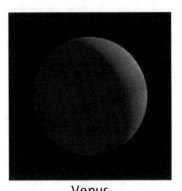

Venus
25th

Mercury

Date	Con	R.A.	Dec	Mag	Diam	Ill.	Elong	AM/PM	Close To
21st	Vir	11h 58m	1° 48'	-1.6	5''	100%	2° E	NV	
23rd	Vir	12h 11m	0° 12'	-1.5	5''	100%	2° E	NV	
25th	Vir	12h 24m	-1° 21'	-1.3	5''	99%	4° E	NV	
27th	Vir	12h 36m	-2° 54'	-1.1	5''	99%	5° E	NV	
29th	Vir	12h 49m	-4° 24'	-1.0	5''	98%	7° E	NV	Spica

Venus

Date	Con	R.A.	Dec	Mag	Diam	Ill	Elong	AM/PM	Close To
21st	Vir	14h 12m	-19° 24'	-4.6	40"	26%	39° E	PM	
23rd	Vir	14h 15m	-19° 54'	-4.6	41"	24%	38° E	PM	
25th	Vir	14h 18m	-20° 24'	-4.6	42"	22%	37° E	PM	
27th	Vir	14h 20m	-20° 48'	-4.6	44"	20%	36° E	PM	
29th	Vir	14h 22m	-21° 9'	-4.6	45"	19%	34° E	PM	

Mars and the Outer Planets

Mars
25th

Jupiter
25th

Saturn
25th

Mars

Date	Con	R.A.	Dec	Mag	Diam	Ill	Elong	AM/PM	Close To
20th	Cap	20h 23m	-24° 3'	-1.6	18"	90%	125° E	PM	Moon
25th	Cap	20h 29m	-23° 18'	-1.5	17"	89%	122° E	PM	
30th	Cap	20h 37m	-22° 36'	-1.3	16"	89%	118° E	PM	

The Outer Planets

Planet	Date	Con	R.A.	Dec	Mag	Diam	Elong	AM/PM	Close To
Jupiter	25th	Lib	15h 15m	17° 18'	-1.8	33"	49° E	PM	
Saturn	25th	Sgr	18h 12m	-22° 45'	0.5	17"	91° E	PM	
Uranus	25th	Ari	01h 59m	11° 30'	5.7	4"	151° W	AM	
Neptune	25th	Aqr	23h 05m	-7° 0'	7.8	2"	162° E	PM	

Events

Date	Time (UT)	Event
21st	01:38	Mercury is at superior conjunction with the Sun. (Not visible.)
23rd	01:55	Autumnal equinox. Autumn begins in the northern hemisphere, spring in the south.
25th	02:53	Full Moon. (All night.)
	04:11	Venus reaches maximum brightness. (Evening sky.)
27th	08:54	The waning gibbous Moon appears south of Uranus. (Morning sky.)
30th	08:43	The waning gibbous Moon appears north of the bright star Aldebaran. (Taurus, morning sky.)

Planet Locations – September 25th

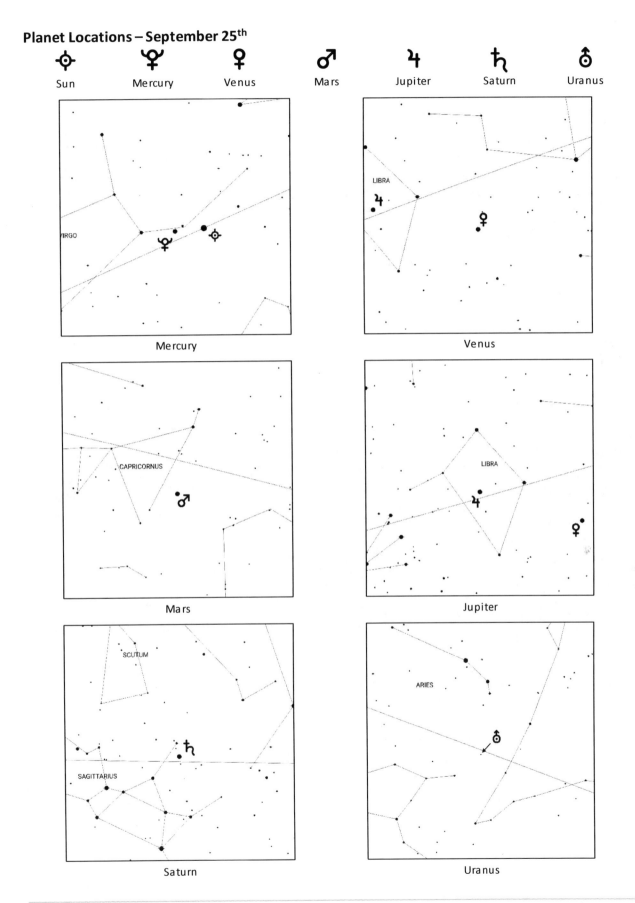

Sun   Mercury   Venus   Mars   Jupiter   Saturn   Uranus

Mercury

Venus

Mars

Jupiter

Saturn

Uranus

October 1st to 10th

The Moon

| 1st | 3rd | 5th | 7th | 9th |

Date	Con	R.A.	Dec	Mag	Diam	Ill.	Elong	Phase		Close To
1st	Tau	05h 45m	20° 18'	-10.2	32'	60%	102° W	LQ		
2nd	Gem	06h 45m	21° 3'	-9.8	32'	49%	89° W	LQ		
3rd	Gem	07h 45m	20° 23'	-9.2	33'	38%	76° W	LQ		
4th	Cnc	08h 45m	18° 30'	-8.6	33'	27%	62° W	-Cr		Praesepe
5th	Leo	09h 44m	15° 24'	-7.9	33'	17%	49° W	-Cr		Regulus
6th	Leo	10h 42m	11° 21'	-7.1	33'	9%	36° W	NM		Regulus
7th	Leo	11h 37m	6° 36'	-6.2	33'	4%	22° W	NM		
8th	Vir	12h 31m	1° 33'	-5.2	33'	1%	10° W	NM		
9th	Vir	13h 24m	-3° 30'	-4.9	33'	0%	7° E	NM		Mercury, Spica
10th	Vir	14h 16m	-8° 18'	-5.8	32'	3%	18° E	NM		Mercury, Venus

Mercury and Venus

| Mercury | Mercury | Venus |
| 3rd | 7th | 5th |

Mercury

Date	Con	R.A.	Dec	Mag	Diam	Ill.	Elong	AM/PM		Close To
1st	Vir	13h 01m	-5° 54'	-0.9	5''	97%	8° E	NV		Spica
3rd	Vir	13h 13m	-7° 24'	-0.7	5''	97%	9° E	NV		Spica
5th	Vir	13h 25m	-8° 48'	-0.6	5''	96%	11° E	NV		Spica
7th	Vir	13h 36m	-10° 12'	-0.6	5''	95%	12° E	NV		Spica
9th	Vir	13h 48m	-11° 33'	-0.5	5''	94%	13° E	NV		Moon, Venus, Spica

Venus

Date	Con	R.A.	Dec	Mag	Diam	Ill	Elong	AM/PM	Close To
1st	Lib	14h 23m	-21° 27'	-4.5	47"	17%	33° E	PM	
3rd	Lib	14h 24m	-21° 42'	-4.5	48"	15%	31° E	PM	
5th	Lib	14h 24m	-21° 54'	-4.5	50"	13%	30° E	PM	
7th	Lib	14h 24m	-22° 0'	-4.5	51"	11%	28° E	PM	
9th	Lib	14h 23m	-22° 0'	-4.5	53"	10%	25° E	PM	Mercury

Mars and the Outer Planets

Mars
5th

Jupiter
5th

Saturn
5th

Mars

Date	Con	R.A.	Dec	Mag	Diam	Ill	Elong	AM/PM	Close To
1st	Cap	20h 38m	-22° 27'	-1.3	16"	88%	118° E	PM	
5th	Cap	20h 45m	-21° 48'	-1.2	15"	88%	116° E	PM	
10th	Cap	20h 54m	-20° 57'	-1.1	14"	87%	113° E	PM	

The Outer Planets

Planet	Date	Con	R.A.	Dec	Mag	Diam	Elong	AM/PM	Close To
Jupiter	5th	Lib	15h 23m	-17° 45'	-1.8	32"	41° E	PM	
Saturn	5th	Sgr	18h 14m	-22° 45'	0.5	16"	81° E	PM	
Uranus	5th	Ari	01h 57m	11° 24'	5.7	4"	161° W	AM	
Neptune	5th	Aqr	23h 04m	-7° 3'	7.8	2"	152° E	PM	

Events

Date	Time (UT)	Event
2nd	09:46	Last Quarter Moon. (Morning sky.)
4th	09:55	Venus is stationary prior to beginning retrograde motion. (Evening sky.)
	11:09	The waning crescent Moon appears south of the Praesepe star cluster. (Cancer, morning sky.)
5th	21:44	The waning crescent Moon appears north of the bright star Regulus. (Leo, morning sky.)
6th	N/A	Good opportunity to see Earthshine on the waning crescent Moon. (Morning sky.)
8th	N/A	The Draconid meteors reach their peak. (ZHR: Variable. All night but best seen in the morning sky.)
9th	03:48	New Moon. (Not visible.)

Planet Locations – October 5th

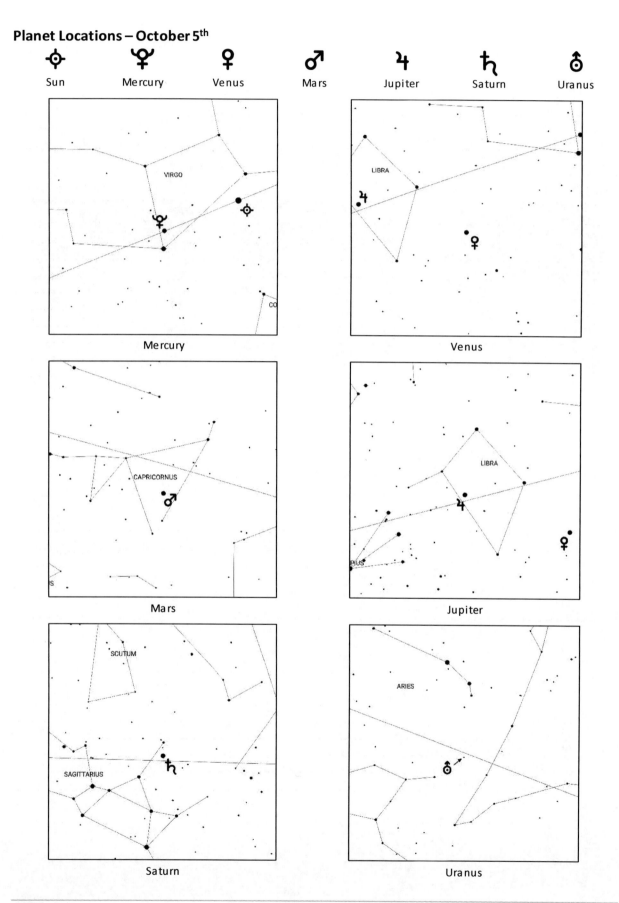

⊕	☿	♀	♂	♃	♄	⛢
Sun	Mercury	Venus	Mars	Jupiter	Saturn	Uranus

Mercury

Venus

Mars

Jupiter

Saturn

Uranus

The Moon

11th　　13th　　15th　　17th　　19th

Date	Con	R.A.	Dec	Mag	Diam	Ill.	Elong	Phase		Close To
11th	Lib	15h 08m	-12° 33'	-6.7	32'	7%	30° E	NM		Venus, Jupiter
12th	Sco	16h 00m	-16° 3'	-7.5	31'	13%	42° E	+Cr		Jupiter, Antares
13th	Oph	16h 52m	-18° 42'	-8.2	31'	21%	54° E	+Cr		Antares
14th	Sgr	17h 44m	-20° 24'	-8.7	30'	29%	65° E	+Cr		Saturn
15th	Sgr	18h 36m	-21° 6'	-9.2	30'	38%	76° E	FQ		Saturn
16th	Sgr	19h 27m	-20° 51'	-9.7	30'	48%	87° E	FQ		
17th	Cap	20h 18m	-19° 39'	-10.1	30'	57%	98° E	FQ		Mars
18th	Cap	21h 07m	-17° 36'	-10.5	29'	66%	109° E	+G		Mars
19th	Cap	21h 55m	-14° 48'	-10.8	30'	75%	120° E	+G		Mars
20th	Aqr	22h 43m	-11° 21'	-11.2	30'	83%	131° E	+G		Neptune

Mercury and Venus

Mercury 13th　　Mercury 17th　　Venus 15th

Mercury

Date	Con	R.A.	Dec	Mag	Diam	Ill.	Elong	AM/PM		Close To
11th	Vir	14h 00m	-12° 51'	-0.4	5''	93%	14° E	NV		Venus, Spica
13th	Vir	14h 11m	-14° 6'	-0.4	5''	91%	15° E	NV		Venus
15th	Lib	14h 22m	-15° 18'	-0.3	5''	90%	16° E	PM		Venus
17th	Lib	14h 34m	-16° 27'	-0.3	5''	89%	17° E	PM		Venus
19th	Lib	14h 45m	-17° 33'	-0.3	5''	87%	18° E	PM		Venus

Venus

Date	Con	R.A.	Dec	Mag	Diam	Ill	Elong	AM/PM	Close To
11th	Vir	14h 21m	-21° 54'	-4.4	54''	8%	23° E	PM	Moon, Mercury
13th	Vir	14h 19m	-21° 45'	-4.4	56''	6%	21° E	PM	Mercury
15th	Vir	14h 16m	-21° 30'	-4.3	57''	5%	18° E	PM	Mercury
17th	Vir	14h 13m	-21° 9'	-4.3	58''	4%	16° E	PM	Mercury
19th	Vir	14h 09m	-20° 42'	-4.2	59''	3%	13° E	NV	Mercury

Mars and the Outer Planets

Mars
15th

Jupiter
15th

Saturn
15th

Mars

Date	Con	R.A.	Dec	Mag	Diam	Ill	Elong	AM/PM	Close To
10th	Cap	20h 54m	-20° 57'	-1.1	14''	87%	113° E	PM	
15th	Cap	21h 03m	-20° 3'	-1.0	14''	87%	110° E	PM	
20th	Cap	21h 13m	-19° 3'	-0.9	13''	87%	108° E	PM	

The Outer Planets

Planet	Date	Con	R.A.	Dec	Mag	Diam	Elong	AM/PM	Close To
Jupiter	15th	Lib	15h 31m	-18° 18'	-1.8	32''	33° E	PM	
Saturn	15th	Sgr	18h 16m	-22° 45'	0.6	16''	72° E	PM	Moon
Uranus	15th	Ari	01h 56m	11° 15'	5.7	4''	171° W	AM	
Neptune	15th	Aqr	23h 03m	-7° 9'	7.8	2''	142° E	PM	

Events

Date	Time (UT)	Event
11th	22:44	The waxing crescent Moon appears north of Jupiter. (Evening sky.)
12th	N/A	Good opportunity to see Earthshine on the waxing crescent Moon. (Evening sky.)
13th	02:04	The waxing crescent Moon appears north of the bright star Antares. (Scorpius, evening sky.)
14th	15:06	Mercury appears 6.8° north of Venus. (Evening sky.)
15th	03:25	The waxing crescent Moon appears north of Saturn. (Evening sky.)
16th	07:48	The almost first quarter Moon appears north of dwarf planet Pluto. (Evening sky.)
	18:03	First Quarter Moon. (Evening sky.)
18th	11:23	The waxing gibbous Moon appears north of Mars. (Evening sky.)
20th	23:18	The waxing gibbous Moon appears south of Neptune. (Evening sky.)

Planet Locations – October 15th

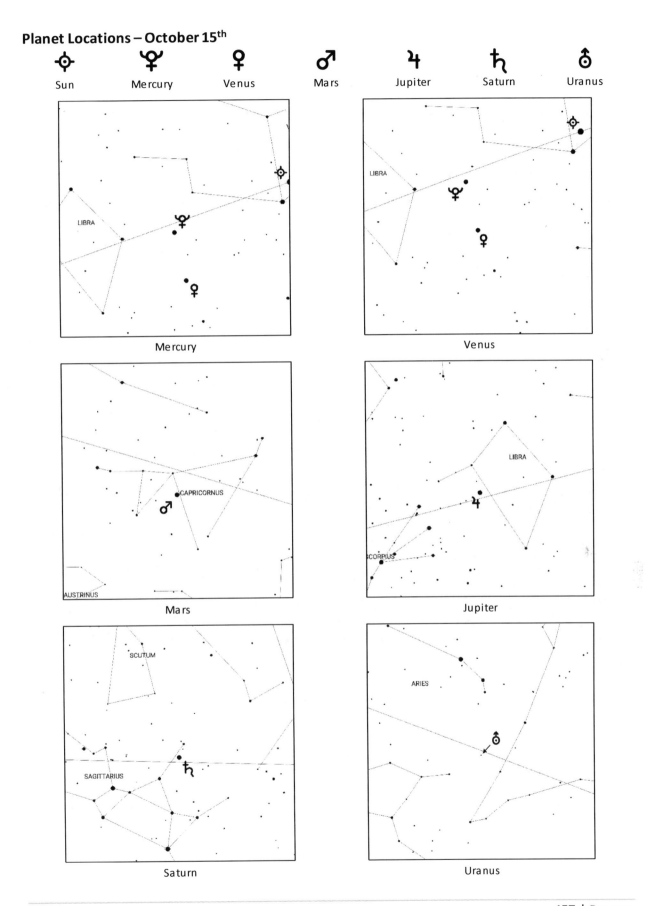

October 21ˢᵗ – 31ˢᵗ

The Moon

| 21ˢᵗ | 23ʳᵈ | 25ᵗʰ | 27ᵗʰ | 29ᵗʰ | 31ˢᵗ |

Date	Con	R.A.	Dec	Mag	Diam	Ill.	Elong	Phase	Close To
21st	Aqr	23h 30m	-7° 21'	-11.5	30'	89%	142° E	+G	
22nd	Psc	00h 17m	-2° 57'	-11.9	30'	95%	153° E	FM	
23rd	Cet	01h 04m	1° 39'	-12.2	30'	98%	165° E	FM	
24th	Psc	01h 53m	6° 18'	-12.5	31'	100%	174° E	FM	Uranus
25th	Ari	02h 44m	10° 45'	-12.3	31'	99%	169° W	FM	
26th	Tau	03h 38m	14° 42'	-12.0	31'	96%	157° W	FM	Pleiades, Hyades
27th	Tau	04h 33m	17° 57'	-11.6	31'	91%	144° W	-G	Pleiades, Hyades, Aldebaran
28th	Tau	05h 32m	20° 12'	-11.2	32'	83%	132° W	-G	
29th	Gem	06h 31m	21° 15'	-10.8	32'	74%	119° W	-G	
30th	Gem	07h 32m	20° 57'	-10.4	32'	64%	106° W	LQ	
31st	Cnc	08h 32m	19° 18'	-9.9	32'	52%	93° W	LQ	Praesepe

Mercury and Venus

| Mercury | Mercury | Venus |
| 23ʳᵈ | 27ᵗʰ | 25ᵗʰ |

Mercury

Date	Con	R.A.	Dec	Mag	Diam	Ill.	Elong	AM/PM	Close To
21st	Lib	14h 56m	-18° 36'	-0.2	5''	86%	19° E	PM	Jupiter
23rd	Lib	15h 07m	-19° 33'	-0.2	5''	84%	20° E	PM	Jupiter
25th	Lib	15h 18m	-20° 27'	-0.2	6''	82%	21° E	PM	Jupiter
27th	Lib	15h 29m	-21° 18'	-0.2	6''	80%	21° E	PM	Jupiter
29th	Lib	15h 40m	-22° 3'	-0.2	6''	77%	22° E	PM	Jupiter
31st	Sco	15h 51m	-22° 45'	-0.2	6''	74%	23° E	PM	Jupiter, Antares

Venus

Date	Con	R.A.	Dec	Mag	Diam	Ill	Elong	AM/PM	Close To
21st	Vir	14h 05m	-20° 9'	-4.1	60''	2%	11° E	NV	
23rd	Vir	14h 01m	-19° 30'	-4.1	61''	1%	8° E	NV	Spica
25th	Vir	13h 57m	-18° 48'	-4.0	61''	1%	7° E	NV	Spica
27th	Vir	13h 53m	-18° 3'	-4.0	61''	1%	6° W	NV	Spica
29th	Vir	13h 49m	-17° 15'	-4.1	61''	1%	7° W	NV	Spica
31st	Vir	13h 45m	-16° 27'	-4.1	61''	1%	9° W	NV	Spica

Mars and the Outer Planets

Mars
25th

Jupiter
25th

Saturn
25th

Mars

Date	Con	R.A.	Dec	Mag	Diam	Ill	Elong	AM/PM	Close To
20th	Cap	21h 13m	-19° 3'	-0.9	13''	87%	108° E	PM	
25th	Cap	21h 23m	-18° 3'	-0.7	13''	86%	106° E	PM	
30th	Cap	21h 34m	-16° 57'	-0.6	12''	86%	103° E	PM	

The Outer Planets

Planet	Date	Con	R.A.	Dec	Mag	Diam	Elong	AM/PM	Close To
Jupiter	25th	Lib	15h 39m	-18° 48'	-1.7	32''	25° E	PM	Mercury
Saturn	25th	Sgr	18h 19m	-22° 45'	0.6	16''	62° E	PM	
Uranus	25th	Ari	01h 54m	11° 5'	5.7	4''	178° E	PM	Moon
Neptune	25th	Aqr	23h 02m	-7° 15'	7.8	2''	132° E	PM	

Events

Date	Time (UT)	Event
21st	N/A	The Orionid meteors reach their peak. (ZHR: 25. All night but best seen in the morning sky.)
24th	07:25	Uranus is at opposition. (All night.)
	12:18	The almost full Moon appears south of Uranus. (All night.)
	16:46	Full Moon. (All night.)
26th	14:09	Venus is at inferior conjunction with the Sun. (Not visible.)
27th	13:24	The waning gibbous Moon appears north of the bright star Aldebaran. (Taurus, morning sky.)
30th	03:39	Mercury appears 3.3° south of Jupiter. (Evening sky.)
31st	16:08	The almost last quarter Moon appears south of the Praesepe star cluster. (Cancer, morning sky.)
	16:41	Last Quarter Moon. (Morning sky.)

Planet Locations – October 25th

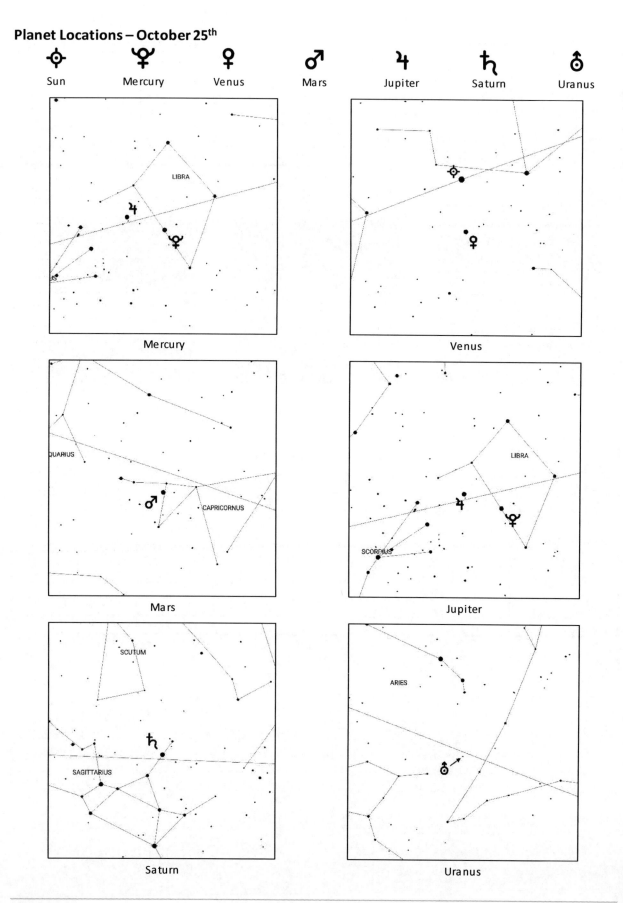

The Moon

1st   3rd   5th   7th   9th

Date	Con	R.A.	Dec	Mag	Diam	Ill.	Elong	Phase	Close To
1st	Leo	09h 30m	16° 30'	-9.4	32'	41%	80° W	LQ	Praesepe, Regulus
2nd	Leo	10h 26m	12° 45'	-8.8	32'	30%	66° W	-Cr	Regulus
3rd	Leo	11h 21m	8° 15'	-8.1	33'	20%	53° W	-Cr	
4th	Vir	12h 13m	3° 21'	-7.4	32'	12%	41° W	NM	
5th	Vir	13h 05m	-1° 39'	-6.5	32'	6%	28° W	NM	Venus, Spica
6th	Vir	13h 56m	-6° 33'	-5.6	32'	2%	15° W	NM	Venus, Spica
7th	Lib	14h 48m	-11° 3'	-4.8	32'	0%	5° W	NM	
8th	Lib	15h 40m	-14° 54'	-5.3	31'	1%	11° E	NM	Mercury, Jupiter, Antares
9th	Oph	16h 32m	-18° 0'	-6.2	31'	4%	22° E	NM	Mercury, Jupiter, Antares
10th	Oph	17h 25m	-20° 6'	-7.0	31'	9%	34° E	NM	

Mercury and Venus

Mercury
3rd

Mercury
7th

Venus
5th

Mercury

Date	Con	R.A.	Dec	Mag	Diam	Ill.	Elong	AM/PM	Close To
1st	Sco	15h 56m	-23° 3'	-0.2	6''	73%	23° E	PM	Jupiter, Antares
3rd	Sco	16h 06m	-23° 36'	-0.2	6''	70%	23° E	PM	Jupiter, Antares
5th	Sco	16h 15m	-24° 3'	-0.2	6''	66%	23° E	PM	Jupiter, Antares
7th	Oph	16h 24m	-24° 24'	-0.2	7''	61%	23° E	PM	Jupiter, Antares
9th	Oph	16h 31m	-24° 39'	-0.1	7''	56%	23° E	PM	Moon, Jupiter, Antares

Venus

Date	Con	R.A.	Dec	Mag	Diam	Ill	Elong	AM/PM		Close To
1st	Vir	13h 43m	-16° 3'	-4.2	61"	2%	11° W	NV		Spica
3rd	Vir	13h 40m	-15° 12'	-4.2	60"	3%	13° W	NV		Spica
5th	Vir	13h 37m	-14° 24'	-4.3	59"	4%	16° W	AM		Moon, Spica
7th	Vir	13h 34m	-13° 39'	-4.4	58"	5%	19° W	AM		Spica
9th	Vir	13h 33m	-12° 57'	-4.4	56"	7%	21° W	AM		Spica

Mars and the Outer Planets

Mars
5th

Jupiter
5th

Saturn
5th

Mars

Date	Con	R.A.	Dec	Mag	Diam	Ill	Elong	AM/PM		Close To
1st	Cap	21h 38m	-16° 30'	-0.6	12"	86%	102° E	PM		
5th	Cap	21h 47m	-16° 36'	-0.5	11"	86%	101° E	PM		
10th	Cap	21h 58m	-14° 24'	-0.4	11"	86%	99° E	PM		

The Outer Planets

Planet	Date	Con	R.A.	Dec	Mag	Diam	Elong	AM/PM		Close To
Jupiter	5th	Lib	15h 49m	-19° 21'	-1.7	31"	16° E	PM		Mercury
Saturn	5th	Sgr	18h 23m	-22° 45'	0.6	16"	52° E	PM		
Uranus	5th	Ari	01h 53m	10° 57'	5.7	4"	167° E	PM		
Neptune	5th	Aqr	23h 02m	-7° 18'	7.9	2"	121° E	PM		

Events

Date	Time (UT)	Event
2nd	02:38	The just-past last quarter Moon appears north of the bright star Regulus. (Leo, morning sky.)
4th	N/A	Good opportunity to see Earthshine on the waning crescent Moon. (Morning sky.)
5th	21:42	The waning crescent Moon appears north of the bright star Spica. (Virgo, morning sky.)
6th	15:18	Mercury is at greatest eastern elongation. (Evening sky.)
7th	16:03	New Moon. (Not visible.)
8th	22:34	Mercury appears 1.8° north of the bright star Antares. (Scorpius, evening sky.)
9th	09:04	The waxing crescent Moon appears north of the bright star Antares. (Scorpius, evening sky.)
	10:11	The waxing crescent Moon appears north of Mercury. (Evening sky.)

Planet Locations – November 5th

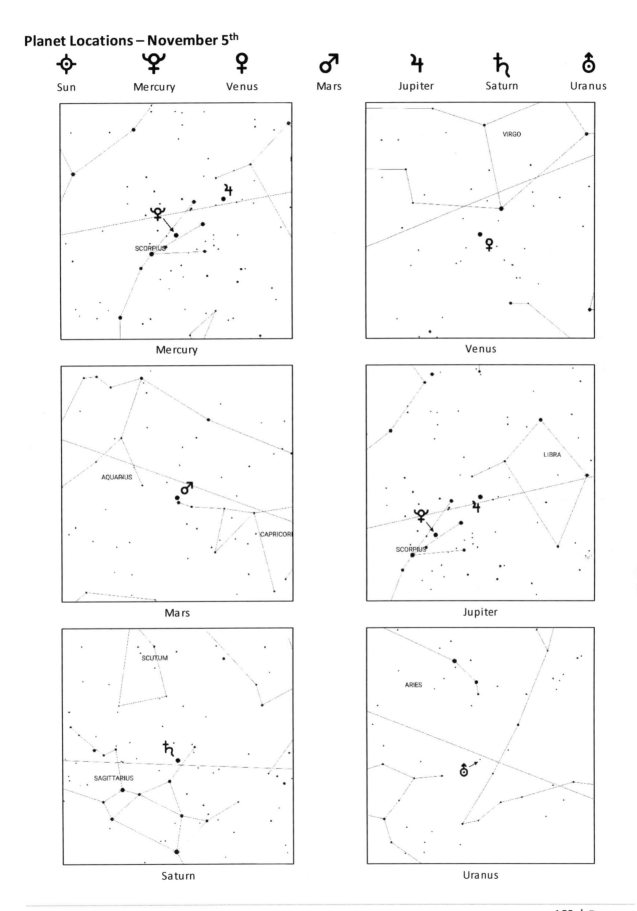

November 11th – 20th

The Moon

| 11th | 13th | 15th | 17th | 19th |

Date	Con	R.A.	Dec	Mag	Diam	Ill.	Elong	Phase		Close To
11th	Sgr	18h 18m	-21° 12'	-7.7	30'	15%	45° E	+Cr		Saturn
12th	Sgr	19h 10m	-21° 18'	-8.3	30'	22%	56° E	+Cr		Saturn
13th	Sgr	20h 01m	-20° 27'	-8.8	30'	31%	67° E	+Cr		
14th	Cap	20h 51m	-18° 39'	-9.3	30'	40%	78° E	FQ		
15th	Cap	20h 39m	-16° 6'	-9.8	30'	49%	89° E	FQ		
16th	Aqr	22h 26m	-12° 51'	-10.2	30'	58%	100° E	FQ		Mars
17th	Aqr	23h 13m	-9° 3'	-10.5	30'	68%	111° E	+G		Neptune
18th	Psc	23h 59m	-4° 48'	-10.9	30'	76%	122° E	+G		
19th	Cet	00h 46m	0° 15'	-11.3	30'	84%	133° E	+G		
20th	Psc	01h 35m	4° 27'	-11.6	31'	91%	145° E	+G		Uranus

Mercury and Venus

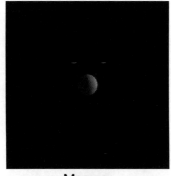  

Mercury	Mercury	Venus
13th	17th	15th

Mercury

Date	Con	R.A.	Dec	Mag	Diam	Ill.	Elong	AM/PM		Close To
11th	Oph	16h 38m	-24° 48'	-0.1	7''	51%	23° E	PM		Antares
13th	Oph	16h 43m	-24° 48'	0.1	8''	44%	22° E	PM		Antares
15th	Oph	16h 46m	-24° 39'	0.3	8''	37%	20° E	PM		Antares
17th	Oph	16h 47m	-24° 21'	0.6	8''	29%	19° E	PM		Antares
19th	Oph	16h 45m	-23° 54'	1.0	9''	21%	16° E	PM		Antares

Venus

Date	Con	R.A.	Dec	Mag	Diam	Ill	Elong	AM/PM	Close To
11th	Vir	13h 32m	-12° 18'	-4.5	55"	8%	23° W	AM	Spica
13th	Vir	13h 31m	-11° 42'	-4.5	54"	10%	26° W	AM	Spica
15th	Vir	13h 31m	-11° 12'	-4.5	52"	12%	28° W	AM	Spica
17th	Vir	13h 32m	-10° 48'	-4.6	51"	13%	30° W	AM	Spica
19th	Vir	13h 33m	-10° 27'	-4.6	49"	15%	32° W	AM	Spica

Mars and the Outer Planets

Mars
15th

Jupiter
15th

Saturn
15th

Mars

Date	Con	R.A.	Dec	Mag	Diam	Ill	Elong	AM/PM	Close To
10th	Cap	21h 58m	-14° 24'	-0.4	11"	86%	99° E	PM	
15th	Aqr	22h 10m	-13° 9'	-0.3	11"	86%	97° E	PM	
20th	Aqr	22h 21m	-11° 54'	-0.2	10"	86%	95° E	PM	

The Outer Planets

Planet	Date	Con	R.A.	Dec	Mag	Diam	Elong	AM/PM	Close To
Jupiter	15th	Lib	15h 58m	-19° 48'	-1.7	31"	9° E	NV	Antares
Saturn	15th	Sgr	18h 27m	-22° 45'	0.6	15"	43° E	PM	
Uranus	15th	Ari	01h 51m	10° 51'	5.7	4"	157° E	PM	
Neptune	15th	Aqr	23h 02m	-7° 18'	7.9	2"	111° E	PM	

Events

Date	Time (UT)	Event
11th	16:05	The waxing crescent Moon appears north of Saturn. (Evening sky.)
	N/A	Good opportunity to see Earthshine on the waxing crescent Moon. (Evening sky.)
12th	19:08	The waxing crescent Moon appears south of Pluto. An occultation may be visible. (Evening sky.)
	20:43	Venus is stationary prior to resuming prograde motion. (Morning sky.)
15th	14:55	First Quarter Moon. (Evening sky.)
16th	05:04	The just-past first quarter Moon appears south of Mars. (Evening sky.)
17th	04:54	Mercury is stationary prior to beginning retrograde motion. (Evening sky.)
	06:28	The waxing gibbous Moon appears south of Neptune. (Evening sky.)
	N/A	The Leonid meteors reach their peak. (ZHR: 15. All night but best seen in the morning sky.)
20th	18:18	The waxing gibbous Moon appears south of Uranus. (Evening sky.)

Planet Locations – November 15th

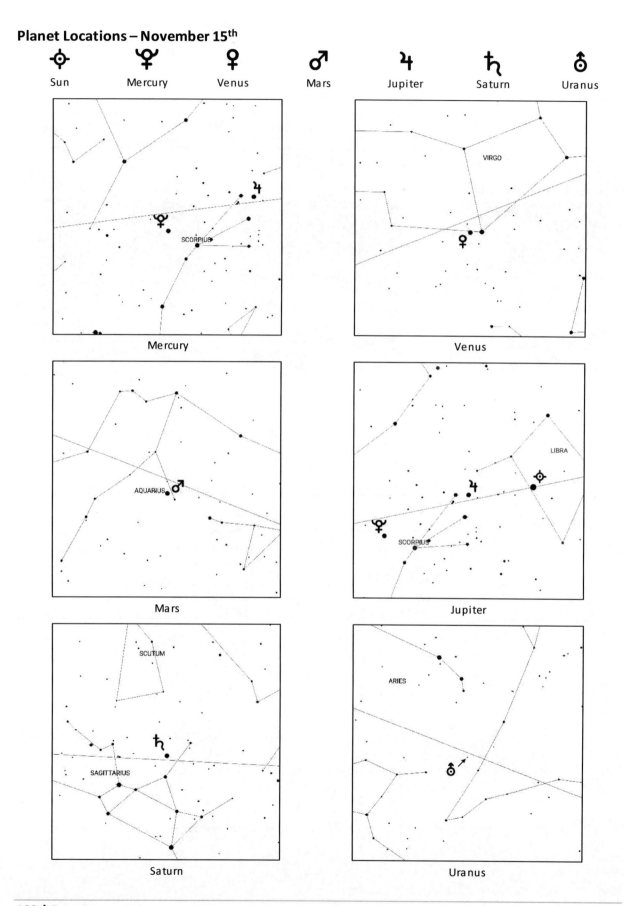

The Moon

21st   23rd   25th   27th   29th

Date	Con	R.A.	Dec	Mag	Diam	Ill.	Elong	Phase	Close To
21st	Cet	02h 25m	9° 3'	-12.0	31'	96%	157° E	FM	
22nd	Ari	03h 18m	13° 21'	-12.4	31'	99%	169° E	FM	Pleiades
23rd	Tau	04h 14m	17° 0'	-12.5	32'	100%	175° W	FM	Pleiades, Hyades, Aldebaran
24th	Tau	05h 12m	19° 45'	-12.2	32'	98%	163° W	FM	Hyades, Aldebaran
25th	Ori	06h 14m	21° 15'	-11.8	32'	93%	150° W	-G	
26th	Gem	07h 16m	21° 21'	-11.4	32'	86%	137° W	-G	
27th	Cnc	08h 18m	20° 3'	-10.9	32'	77%	132° W	-G	Praesepe
28th	Cnc	09h 17m	17° 27'	-10.5	32'	67%	110° W	-G	Praesepe
29th	Leo	10h 14m	13° 51'	-10.1	32'	56%	97° W	LQ	Regulus
30th	Leo	11h 09m	9° 33'	-9.6	32'	45%	84° W	LQ	

Mercury and Venus

Mercury   Mercury   Venus
23rd   27th   25th

Mercury

Date	Con	R.A.	Dec	Mag	Diam	Ill.	Elong	AM/PM	Close To
21st	Oph	16h 41m	-23° 12'	1.7	9''	13%	13° E	NV	Jupiter, Antares
23rd	Oph	16h 33m	-22° 21'	2.7	10''	6%	9° E	NV	Jupiter, Antares
25th	Oph	16h 23m	-21° 18'	4.1	10''	2%	5° E	NV	Jupiter, Antares
27th	Sco	16h 13m	-20° 9'	5.4	10''	0%	1° W	NV	Jupiter, Antares
29th	Lib	16h 02m	-19° 3'	3.8	10''	2%	5° W	NV	Jupiter, Antares

Venus

Date	Con	R.A.	Dec	Mag	Diam	Ill	Elong	AM/PM	Close To
21st	Vir	13h 35m	-10° 12'	-4.6	47''	17%	33° W	AM	Spica
23rd	Vir	13h 37m	-10° 0'	-4.6	46''	19%	35° W	AM	Spica
25th	Vir	13h 40m	-9° 51'	-4.6	45''	21%	36° W	AM	Spica
27th	Vir	13h 43m	-9° 48'	-4.7	43''	23%	38° W	AM	Spica
29th	Vir	13h 47m	-9° 51'	-4.7	42''	24%	39° W	AM	Spica

Mars and the Outer Planets

Mars
25th

Jupiter
25th

Saturn
25th

Mars

Date	Con	R.A.	Dec	Mag	Diam	Ill	Elong	AM/PM	Close To
20th	Aqr	22h 21m	-11° 54'	-0.2	10''	86%	95° E	PM	
25th	Aqr	22h 33m	-10° 33'	-0.1	10''	86%	93° E	PM	
30th	Aqr	22h 45m	-9° 12'	0.0	9''	86%	90° E	PM	Neptune

The Outer Planets

Planet	Date	Con	R.A.	Dec	Mag	Diam	Elong	AM/PM	Close To
Jupiter	25th	Sco	16h 08m	-20° 15'	-1.7	31''	1° E	NV	Mercury, Antares
Saturn	25th	Sgr	18h 31m	-22° 42'	0.6	15''	34° E	PM	
Uranus	25th	Ari	01h 50m	10° 42'	5.7	4''	146° E	PM	
Neptune	25th	Aqr	23h 01m	-7° 18'	7.9	2''	101° E	PM	

Events

Date	Time (UT)	Event
23rd	05:40	Full Moon. (All night.)
	19:37	The full Moon appears north of the bright star Aldebaran. (Taurus, all night.)
25th	05:52	Neptune is stationary prior to resuming prograde motion. (Evening sky.)
26th	10:23	Jupiter is in conjunction with the Sun. (Not visible.)
27th	09:08	Mercury is at inferior conjunction with the Sun. (Not visible.)
29th	11:08	The nearly last quarter Moon appears north of the bright star Regulus. (Leo, morning sky.)
30th	00:20	Last Quarter Moon. (Morning sky.)
	02:16	Venus reaches its maximum brightness. (Morning sky.)

Planet Locations – November 25th

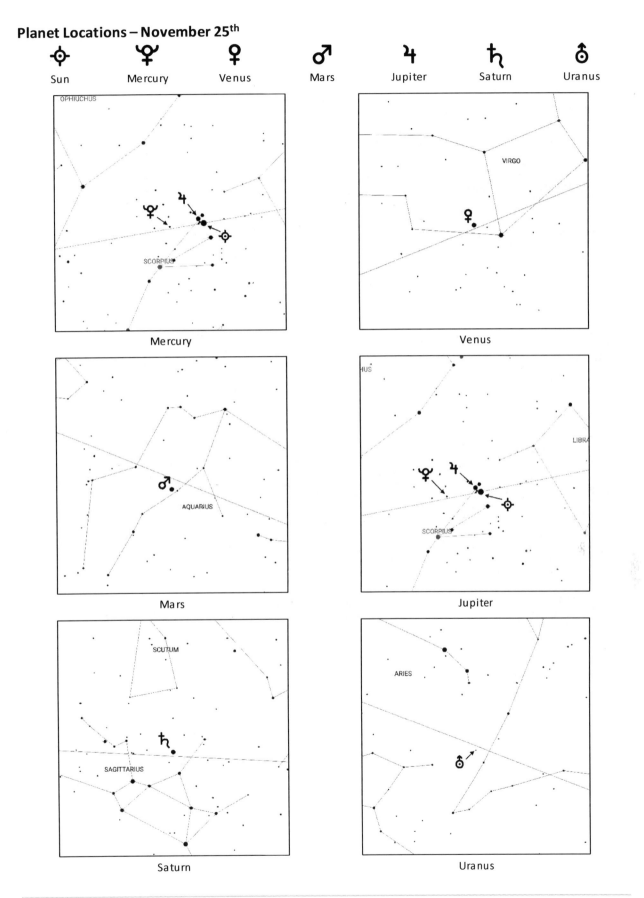

December 1st to 10th

The Moon

1st

3rd

5th

7th

9th

Date	Con	R.A.	Dec	Mag	Diam	Ill.	Elong	Phase	Close To
1st	Vir	12h 01m	4° 45'	-9.0	32'	34%	71° W	-Cr	
2nd	Vir	12h 52m	0° 12'	-8.4	32'	24%	58° W	-Cr	Spica
3rd	Vir	13h 42m	-5° 6'	-7.7	32'	15%	46° W	-Cr	Venus, Spica
4th	Lib	14h 33m	-9° 42'	-7.0	32'	9%	34° W	NM	Venus
5th	Lib	15h 23m	-13° 45'	-6.1	31'	4%	22° W	NM	Mercury
6th	Sco	16h 15m	-17° 6'	-5.2	31'	1%	10° W	NM	Mercury, Jupiter, Antares
7th	Oph	17h 07m	-19° 33'	-4.7	31'	0%	4° E	NM	Jupiter, Antares
8th	Sgr	18h 00m	-21° 3'	-5.5	31'	2%	14° E	NM	Saturn
9th	Sgr	18h 52m	-21° 33'	-6.3	30'	5%	25° E	NM	Saturn
10th	S	19h 44m	-21° 0'	-7.1	30'	10%	36° E	NM	

Mercury and Venus

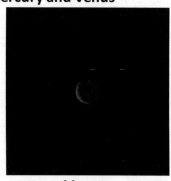

Mercury
3rd

Mercury
7th

Venus
5th

Mercury

Date	Con	R.A.	Dec	Mag	Diam	Ill.	Elong	AM/PM	Close To
1st	Lib	15h 53m	-18° 6'	2.4	9''	8%	10° W	NV	Jupiter, Antares
3rd	Lib	15h 47m	-17° 24'	1.4	9''	15%	13° W	NV	Jupiter
5th	Lib	15h 43m	-17° 0'	0.7	9''	24%	16° W	AM	Moon, Jupiter
7th	Lib	15h 43m	-16° 54'	0.3	8''	33%	18° W	AM	Jupiter
9th	Lib	15h 45m	-17° 3'	0.0	8''	42%	20° W	AM	Jupiter

Venus

Date	Con	R.A.	Dec	Mag	Diam	Ill	Elong	AM/PM		Close To
1st	Vir	13h 51m	-9° 54'	-4.7	40''	26%	40° W	AM		Spica
3rd	Vir	13h 56m	-10° 3'	-4.6	39''	28%	41° W	AM		Moon, Spica
5th	Vir	14h 01m	-10° 12'	-4.6	38''	30%	42° W	AM		Spica
7th	Vir	14h 06m	-10° 27'	-4.6	37''	31%	43° W	AM		
9th	Vir	14h 11m	-10° 42'	-4.6	36''	33%	43° W	AM		

Mars and the Outer Planets

Mars
5th

Jupiter
5th

Saturn
5th

Mars

Date	Con	R.A.	Dec	Mag	Diam	Ill	Elong	AM/PM		Close To
1st	Aqr	22h 47m	-8° 57'	0.0	9''	86%	91° E	PM		Neptune
5th	Aqr	22h 57m	-7° 51'	0.0	9''	86%	89° E	PM		Neptune
10th	Aqr	23h 08m	-6° 27'	0.1	9''	86%	87° E	PM		Neptune

The Outer Planets

Planet	Date	Con	R.A.	Dec	Mag	Diam	Elong	AM/PM		Close To
Jupiter	5th	Sco	16h 17m	-20° 39'	-1.7	31''	7° W	NV		Mercury, Antares
Saturn	5th	Sgr	18h 36m	-22° 39'	0.5	15''	25° E	PM		
Uranus	5th	Ari	01h 49m	10° 36'	5.7	4''	136° E	PM		
Neptune	5th	Aqr	23h 02m	-7° 18'	7.9	2''	90° E	PM		Mars

Events

Date	Time (UT)	Event
3rd	01:59	The waning crescent Moon appears north of the bright star Spica. (Virgo, morning sky.)
	19:30	The waning crescent Moon appears north of Venus. (Morning sky.)
4th	N/A	Good opportunity to see Earthshine on the waning crescent Moon. (Morning sky.)
5th	21:38	The waning crescent Moon appears north of Mercury. (Morning sky.)
6th	20:16	Mercury is stationary prior to resuming prograde motion. (Morning sky.)
7th	07:21	New Moon. (Not visible.)
	14:39	Mars appears 0.0° north of Neptune. (Evening sky.)
9th	04:09	The waxing crescent Moon appears north of Saturn. (Evening sky.)

Planet Locations – December 5th

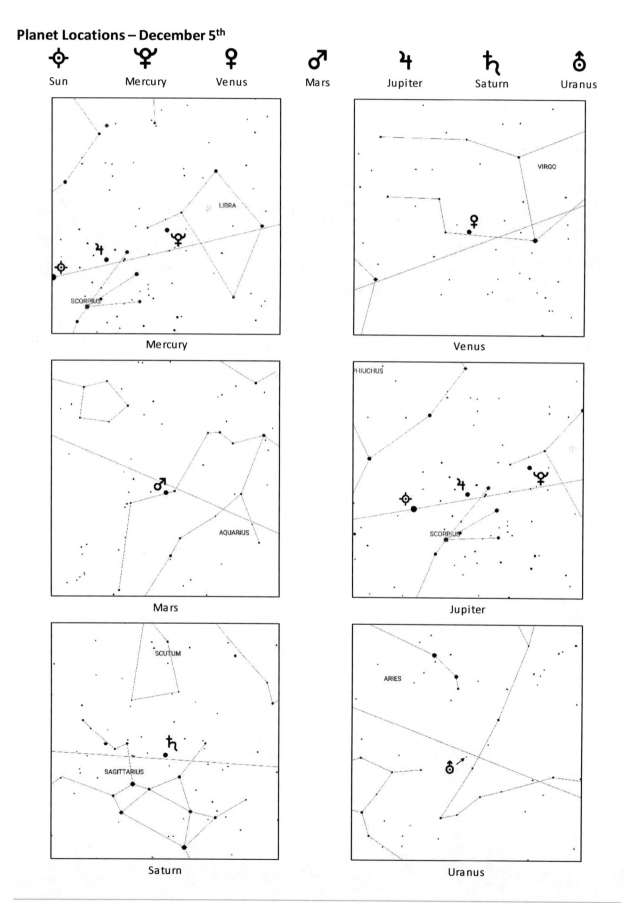

The Moon

| 11th | 13th | 15th | 17th | 19th |

Date	Con	R.A.	Dec	Mag	Diam	Ill.	Elong	Phase	Close To
11th	Cap	20h 35m	-19° 30'	-7.8	30'	16%	47° E	+Cr	
12th	Cap	21h 23m	-17° 12'	-8.4	30'	23%	58° E	+Cr	
13th	Aqr	22h 11m	-14° 12'	-8.9	30'	32%	68° E	+Cr	
14th	Aqr	22h 57m	-10° 36'	-9.4	30'	41%	79° E	FQ	Mars, Neptune
15th	Aqr	23h 43m	-6° 33'	-9.8	30'	50%	90° E	FQ	Mars
16th	Cet	00h 29m	-2° 9'	-10.2	30'	60%	101° E	FQ	
17th	Cet	01h 16m	2° 24'	-10.6	30'	69%	112° E	+G	
18th	Cet	02h 04m	7° 2'	-11.0	31'	79%	125° E	+G	Uranus
19th	Ari	02h 55m	11° 27'	-11.4	31'	87%	137° E	+G	
20th	Tau	03h 49m	15° 27'	-11.7	32'	93%	150° E	+G	Pleiades, Hyades, Aldebaran

Mercury and Venus

| Mercury | Mercury | Venus |
| 13th | 17th | 15th |

Mercury

Date	Con	R.A.	Dec	Mag	Diam	Ill.	Elong	AM/PM	Close To
11th	Lib	15h 49m	-17° 21'	-0.2	7''	50%	21° W	AM	Jupiter
13th	Lib	15h 55m	-17° 51'	-0.3	7''	57%	21° W	AM	Jupiter, Antares
15th	Sco	16h 03m	-18° 24'	-0.4	7''	63%	21° W	AM	Jupiter, Antares
17th	Sco	16h 11m	-19° 3'	-0.4	6''	68%	21° W	AM	Jupiter, Antares
19th	Sco	16h 21m	-19° 42'	-0.4	6''	73%	21° W	AM	Jupiter, Antares

Venus

Date	Con	R.A.	Dec	Mag	Diam	Ill	Elong	AM/PM	Close To
11th	Vir	14h 17m	-11° 0'	-4.6	35''	34%	44° W	AM	
13th	Lib	14h 23m	-11° 18'	-4.6	34''	36%	44° W	AM	
15th	Lib	14h 29m	-11° 42'	-4.6	33''	37%	45° W	AM	
17th	Lib	14h 35m	-12° 6'	-4.6	32''	38%	45° W	AM	
19th	Lib	14h 42m	-12° 30'	-4.6	31''	40%	46° W	AM	

Mars and the Outer Planets

Mars
15th

Jupiter
15th

Saturn
15th

Mars

Date	Con	R.A.	Dec	Mag	Diam	Ill	Elong	AM/PM	Close To
10th	Aqr	23h 08m	-6° 27'	0.1	9''	86%	87° E	PM	Neptune
15th	Aqr	23h 20m	-5° 3'	0.2	8''	87%	86° E	PM	Moon, Neptune
20th	Aqr	23h 32m	-3° 36'	0.3	8''	87%	84° E	PM	

The Outer Planets

Planet	Date	Con	R.A.	Dec	Mag	Diam	Elong	AM/PM	Close To
Jupiter	15th	Oph	16h 26m	-21° 3'	-1.7	31''	15° W	NV	Mercury, Antares
Saturn	15th	Sgr	18h 41m	-22° 36'	0.5	15''	16° E	PM	
Uranus	15th	Ari	01h 48m	10° 33'	5.7	4''	125° E	PM	
Neptune	15th	Aqr	23h 02m	-7° 15'	7.9	2''	80° E	PM	Moon, Mars

Events

Date	Time (UT)	Event
11th	N/A	Good opportunity to see Earthshine on the waxing crescent Moon. (Evening sky.)
13th	N/A	The Geminid meteor shower reaches its peak. (ZHR: 120. All night but best seen in the morning sky.)
14th	12:26	The nearly first quarter Moon appears south of Neptune. (Evening sky.)
15th	01:17	The nearly first quarter Moon appears south of Mars. (Evening sky.)
	11:15	Mercury is at greatest western elongation. (Morning sky.)
	11:50	First Quarter Moon. (Evening sky.)
18th	05:21	The waxing gibbous Moon appears south of Uranus. (Evening sky.)
20th	16:56	Mercury reaches its maximum magnitude. (Morning sky.)

Planet Locations – December 15th

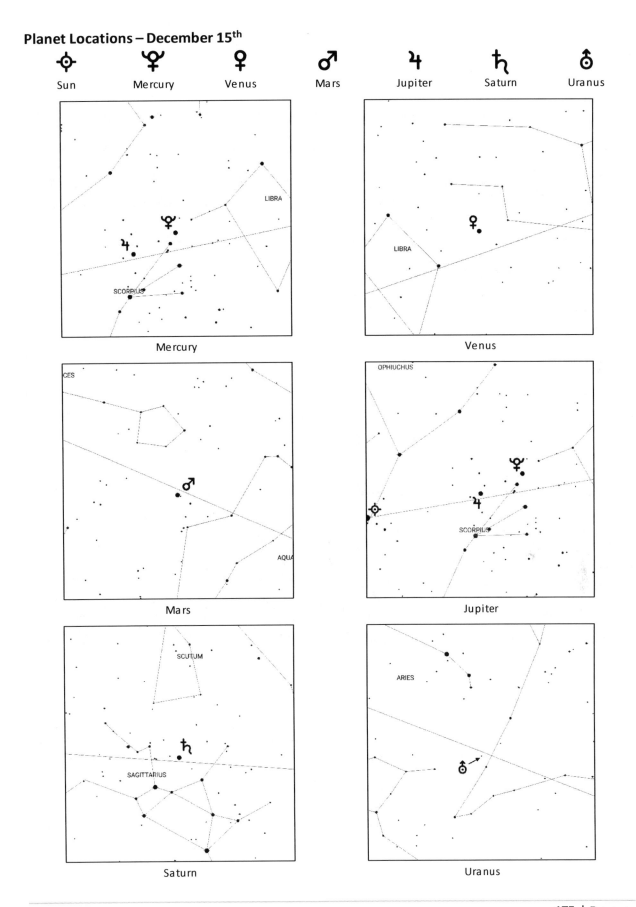

⊙	☿	♀	♂	♃	♄	⛢
Sun	Mercury	Venus	Mars	Jupiter	Saturn	Uranus

Mercury

Venus

Mars

Jupiter

Saturn

Uranus

December 21st – 31st

The Moon

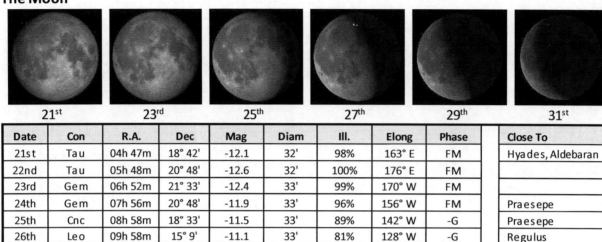

| 21st | 23rd | 25th | 27th | 29th | 31st |

Date	Con	R.A.	Dec	Mag	Diam	Ill.	Elong	Phase	Close To
21st	Tau	04h 47m	18° 42'	-12.1	32'	98%	163° E	FM	Hyades, Aldebaran
22nd	Tau	05h 48m	20° 48'	-12.6	32'	100%	176° E	FM	
23rd	Gem	06h 52m	21° 33'	-12.4	33'	99%	170° W	FM	
24th	Gem	07h 56m	20° 48'	-11.9	33'	96%	156° W	FM	Praesepe
25th	Cnc	08h 58m	18° 33'	-11.5	33'	89%	142° W	-G	Praesepe
26th	Leo	09h 58m	15° 9'	-11.1	33'	81%	128° W	-G	Regulus
27th	Leo	10h 55m	10° 51'	-10.7	32'	71%	115° W	-G	Regulus
28th	Vir	11h 49m	6° 2'	-10.2	32'	60%	102° W	LQ	
29th	Vir	12h 41m	1° 3'	-9.8	32'	49%	89° W	LQ	Spica
30th	Vir	13h 31m	-3° 54'	-9.2	32'	38%	76° W	LQ	Spica
31st	Vir	14h 21m	-8° 33'	-8.7	31'	28%	64° W	-Cr	

Mercury and Venus

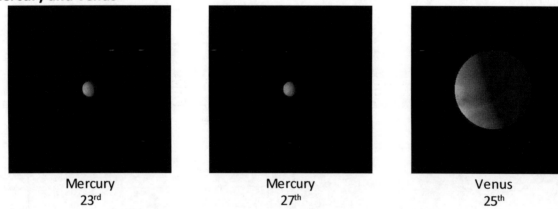

| Mercury | Mercury | Venus |
| 23rd | 27th | 25th |

Mercury

Date	Con	R.A.	Dec	Mag	Diam	Ill.	Elong	AM/PM	Close To
21st	Oph	16h 31m	-20° 21'	-0.4	6''	77%	20° W	AM	Jupiter, Antares
23rd	Oph	16h 42m	-21° 0'	-0.4	6''	80%	20° W	AM	Jupiter, Antares
25th	Oph	16h 54m	-21° 36'	-0.4	6''	83%	19° W	AM	Jupiter, Antares
27th	Oph	17h 06m	-22° 9'	-0.4	5''	85%	18° W	AM	Jupiter, Antares
29th	Oph	17h 18m	-22° 39'	-0.4	5''	87%	18° W	AM	
31st	Oph	17h 30m	-23° 3'	-0.4	5''	89%	17° W	AM	

Venus

Date	Con	R.A.	Dec	Mag	Diam	Ill	Elong	AM/PM		Close To
21st	Lib	14h 49m	-12° 54'	-4.6	30''	41%	46° W	AM		
23rd	Lib	14h 56m	-13° 21'	-4.5	29''	42%	46° W	AM		
25th	Lib	15h 03m	-13° 48'	-4.5	29''	44%	46° W	AM		
27th	Lib	15h 11m	-14° 15'	-4.5	28''	45%	47° W	AM		
29th	Lib	15h 18m	-14° 42'	-4.5	27''	46%	47° W	AM		
31st	Lib	15h 26m	-15° 12'	-4.5	26''	47%	47° W	AM		

Mars and the Outer Planets

Mars
25th

Jupiter
25th

Saturn
25th

Mars

Date	Con	R.A.	Dec	Mag	Diam	Ill	Elong	AM/PM		Close To
20th	Aqr	23h 32m	-3° 36'	0.3	8''	87%	84° E	PM		
25th	Psc	23h 45m	-2° 9'	0.4	8''	87%	82° E	PM		
30th	Psc	23h 57m	-1° 0'	0.4	8''	87%	80° E	PM		

The Outer Planets

Planet	Date	Con	R.A.	Dec	Mag	Diam	Elong	AM/PM	Close To
Jupiter	25th	O	16h 36m	-21° 21'	-1.8	32''	23° W	AM	Mercury, Antares
Saturn	25th	Sgr	18h 46m	-22° 30'	0.5	15''	7° E	NV	
Uranus	25th	Psc	01h 47m	10° 30'	5.7	4''	115° E	PM	
Neptune	25th	Aqr	23h 02m	-7° 12'	7.9	2''	70° E	PM	

Events

Date	Time (UT)	Event
21st	03:03	Mercury appears 6.2° north of the bright star Antares. (Scorpius, morning sky.)
	08:27	The almost full Moon appears north of the bright star Aldebaran. (Taurus, evening sky.)
	14:43	Mercury appears 0.9° north of Jupiter. (Morning sky.)
	22:23	Hibernal solstice. Winter begins in the northern hemisphere, summer in the south.
22nd	17:49	Full Moon. (All night.)
23rd	N/A	The Ursid meteors are at their peak. (ZHR: 10. All night but best in the morning sky.)
26th	16:00	The waning gibbous Moon appears north of the bright star Regulus. (Leo, morning sky.)
29th	09:35	Last Quarter Moon. (Morning sky.)
30th	10:33	The just-past last quarter Moon appears north of the bright star Spica. (Virgo, morning sky.)

Planet Locations – December 25th

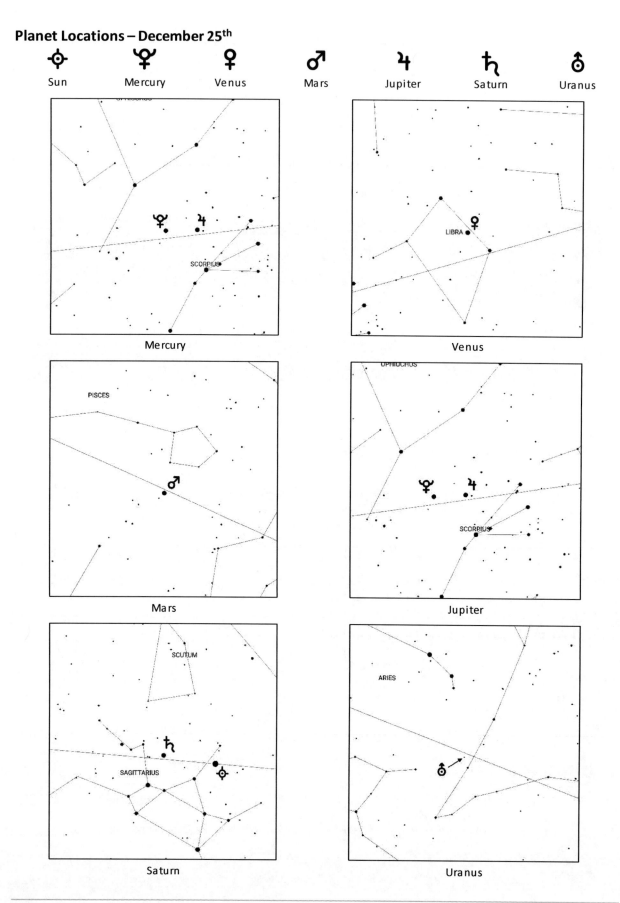

Glossary

Aphelion

The point at which an object is farthest from the Sun. (See also *perihelion*.)

Apogee

The point at which an object is farthest from the Earth. (See also *perigee*.)

Apparent Diameter

Apparent Diameter is the size an object appears in the sky and is measured in degrees, arc-minutes and arc-seconds. If you were to stand facing due north and slowly turn toward east, south, west and then north again, you would be turning 360° (degrees.)

You might therefore think that you can see 360° of sky overhead – in fact, you can only see 180° because the sky is only the visible half a sphere and since you can't see the entire sphere (the ground is in the way,) you can't look 360° in every direction at the sky.

If, however, you were to face due north and look directly overhead at the zenith, this would be 90°. Look down from the zenith to the southern horizon and you would see another 90°, making 180° total. (Incidentally, how high an object appears in the sky is called its *altitude*, but it isn't necessary to know that to use this book.)

To put this into perspective, the Sun and Moon both appear to be about half a degree in diameter but because these are the largest astronomical objects in the sky and everything else appears to be smaller, we need a more convenient (and accurate) measurement.

A degree then is broken up into sixty arc-minutes. Therefore, because the Sun and Moon both appear to be about half a degree, we say their apparent diameter is 30' (arc-minutes.)

The planets and asteroids are even smaller, so we break each arc-minute up into sixty again, thereby creating arc-seconds. A planets' apparent diameter will greatly depend on its actual size and its distance from Earth.

For example, although the planet Venus is slightly smaller than the Earth, it is also the closest world to our own. So at its closest (inferior conjunction) it can have an apparent diameter of 66.01" – in other words, 1' 01" (one arc-minute and one arc-second.) Despite Jupiter being large enough to swallow all the other planets within it, it is much further away – therefore, at its best, it is only able to reach 50.12" (fifty arc-seconds). Neptune is the fourth largest planet but is also the most distant and barely manages to reach 2.37" (arc-seconds.)

Even through a small telescope observers can easily see all of the planets as discs; however, how large the planet appears and the details seen will vary greatly, depending upon the equipment used and the apparent diameter of the planet itself. The dark bands of Jupiter's' atmosphere are easily visible in almost any sized 'scope, whereas Uranus and Neptune typically only show tiny discs in a small to medium sized telescope. Under low power those distant worlds can easily be mistaken for stars.

Only the apparent diameters of the Moon and planets are noted in this book; the dwarf planets Pluto and Ceres, along with the asteroids Juno, Pallas and Vesta and all the bright stars mentioned, only appear as points of light through amateur instruments and their apparent diameter is therefore negligible.

To help put this into perspective, the image below depicts the average apparent size of the planets in comparison with one another.

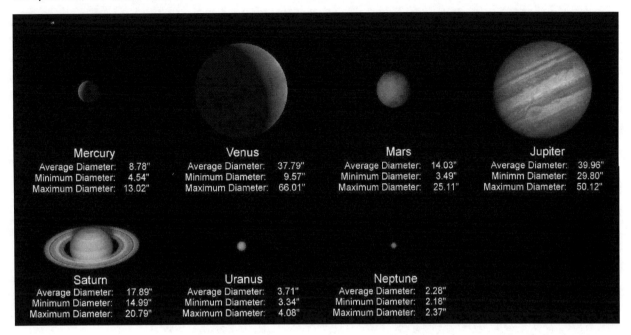

Mercury		Venus		Mars		Jupiter	
Average Diameter:	8.78"	Average Diameter:	37.79"	Average Diameter:	14.03"	Average Diameter:	39.96"
Minimum Diameter:	4.54"	Minimum Diameter:	9.57"	Minimum Diameter:	3.49"	Minimm Diameter:	29.80"
Maximum Diameter:	13.02"	Maximum Diameter:	66.01"	Maximum Diameter:	25.11"	Maximum Diameter:	50.12"

Saturn		Uranus		Neptune	
Average Diameter:	17.89"	Average Diameter:	3.71"	Average Diameter:	2.28"
Minimum Diameter:	14.99"	Minimum Diameter:	3.34"	Minimum Diameter:	2.18"
Maximum Diameter:	20.79"	Maximum Diameter:	4.08"	Maximum Diameter:	2.37"

Asterism

An asterism is a recognizable pattern of stars within a constellation. For example, the "backwards question mark" depicting the head of Leo the Lion, or the seven stars of the Big Dipper (aka, the Plough) in the much-larger constellation of Ursa Major, the Great Bear.

AU – Astronomical Unit

An astronomical unit is basically the mean distance of the Earth to the Sun and is the standard measurement of distance within the solar system. One astronomical unit is equivalent to almost 150 million kilometers (specifically, 149,597,870 km) or roughly 92.96 million miles.

Conjunction

Astronomically, this is a fairly vague term. It basically means any situation when two bodies appear close to one another in the sky. However, there is no officially recognized separation limit that would clearly define when a conjunction is taking place. An object in conjunction with the Sun is never visible because the light from the Sun is too over-powering and the object will be lost in the glare. (See also *inferior conjunction, opposition* and *superior conjunction.*)

Culminate

Culmination occurs when an object is at its highest point in the sky. For most objects, this means it is due south, but this will depend on the object and your latitude. For example, some stars or objects may be directly overhead when they culminate. (In fact, if you faced due south and followed an invisible line between due south and due north, your gaze would pass overhead and any object culminating could also be in sight.)

Earthshine

Earthshine is when the Moon is a crescent but you can see the "dark side" of the Moon too – so you can see the whole Moon in the sky. This happens when light is reflected from the daylight side of the Earth and illuminates the unlit portion of the Moons' surface. It can make for a beautiful sight in the twilight, especially when the Moon is close to a bright star or planet.

Left: An example of Earthshine on the waxing crescent Moon. Light is reflected from the Earth, causing the "dark side" of the Moon to be visible. In this image, the unlit portion of the Moon has been lightened to be more apparent. Image by Steve Jurvetson and used under the Creative Commons Attribution 2.0 Generic license.

Ecliptic

The approximate path the Sun, Moon and planets appear to follow across the sky. The ecliptic crosses the traditional twelve signs of the zodiac as well as briefly passing through other constellations, such as Ophiuchus and Orion. The ecliptic is depicted as a pale blue line in the images used throughout this book.

Elongation

Elongation is how far to the east or west an object appears in relation to another object (most usually in relation to the Sun.) If Mercury or Venus is at eastern elongation, it will appear in the evening sky. If Mercury or Venus is at western elongation, it will appear in the pre-dawn sky. (It is worth noting that there is no guarantee the planet will be visible – it will also depend upon the time of year and the observers' latitude.)

Gibbous

The Moon is said to be gibbous between the half phases (first and last quarter) and full Moon. It's hard to describe the shape – it's not a half Moon, but it's not completely circular either. The inner planets Mercury and Venus can also show a gibbous phase. (See also *illumination*.)

Globular Star Cluster

A globular star cluster is, quite literally and simply, a sphere of thousands of stars. Globular clusters appear as faint, misty balls of grey light against the night, and although they all require at least a pair of binoculars to be seen, many can be resolved into their individual stars through a telescope. The best (and most famous) example in the northern hemisphere is M13, the Great Hercules Cluster (see image below and also *open star cluster*.)

Left: M13, the Great Hercules Cluster. Photo taken by the author using Slooh.

Illumination

Simply how much of an object's visible surface is lit. For example, when the Moon is new, none of the lit surface is visible, so the Moon is 0% illuminated. At half phase (first or last quarter) half the lit surface is visible, so the Moon is 50% illuminated. At full Moon, the entire lit surface is visible, so the Moon is 100% illuminated. The planets Mercury, Venus and Mars can also show phases (Mars, being an outer planet, is more limited) and so their illumination will also change over time. (See also *gibbous*.)

Inferior Conjunction

Inferior conjunction occurs when either Mercury or Venus are directly between the Earth and the Sun. Because the planet will appear so close to the Sun in the sky, it will not be visible from Earth. Mercury and Venus are the only two planets that can go through inferior conjunction because only these two worlds orbit closer to the Sun than the Earth. (See also *conjunction* and *superior conjunction*.)

Magnitude

An object's magnitude is simply a measurement of its brightness. The ancient Greeks created a system where the brightest stars were given a magnitude of 1 and the faintest were magnitude 6. Since that time, astronomers have refined the system and increased its accuracy, but as a result, the magnitude range has increased dramatically.

For example, there are some objects that are brighter than zero and therefore have a negative magnitude. Sirius, the brightest star in the sky, has a magnitude of -1.47. All of the naked eye planets – Mercury, Venus, Mars, Jupiter and Saturn – can all have negative magnitudes. The other two planets, Uranus and Neptune, dwarf planets and asteroids are all more than magnitude five.

(The bright star Vega, in the constellation Lyra, is used as the standard reference point. It has a magnitude of 0.0.)

The naked eye can, theoretically, see objects up to magnitude six, but this greatly depends upon the observer's vision and the conditions of the night sky. Most people can see up to around magnitude five under clear, dark, rural skies. Light pollution is so bad in many towns and cities that, even in the suburbs, it is often difficult to see anything fainter than magnitude 3 or 4 at best.

However, the Moon, Mercury, Venus, Mars, Jupiter and Saturn should easily be visible to anyone, anywhere, assuming that the object is not too close to the Sun. A planets' magnitude will vary, depending upon how close it is to the Earth, how large it appears in the sky and how much of its lit surface is visible. (See also *apparent diameter* and *illumination*.)

Constellations can be problematic, depending upon the brightness of the stars that form the constellation itself.

Meteors are best observed from rural skies but the bright stars mentioned, as well as the Pleiades and Hyades star clusters can be seen from the suburbs.

Almost everything else is best observed under rural skies and/or with binoculars or a telescope.

The magnitude ranges of the solar system objects mentioned in this book are detailed below. (The Sun is, on average, about magnitude -26.74)

Object	Minimum Magnitude (Faintest)	Maximum Magnitude (Brightest)
Moon	-2.5 (New)	-12.9 (Full)
Mercury	5.73	-2.45
Venus	-3.82	-4.89
Mars	1.84	-2.91
Jupiter	-1.61	-2.94
Saturn	1.47	-0.49
Uranus	5.95	5.32
Neptune	8.02	7.78
Pluto	16.3	13.65
Dwarf planet Ceres	9.34	6.64
Asteroid 2 Pallas	10.65	6.49
Asteroid 3 Juno	11.55	7.4
Asteroid 4 Vesta	8.48	5.1

Occultation

When one object completely covers another. Many occultations involve the Moon occulting a star or, sometimes, a planet, but on occasion, a planet may be seen to occult a star. On very rare occasions, one planet may occult another.

Open Star Cluster

An open star cluster is one where the stars appear to be loosely scattered against the night. Unlike globular clusters, their member stars are usually quite young and only number a couple of hundred at most. Several open clusters can be seen with the naked eye (for example, M44 - the Praesepe - in Cancer and the Hyades, which forms the V shaped asterism in the constellation Taurus.) The most famous example of an open cluster is M45, the Pleiades, a naked eye cluster (also in Taurus) that is easily seen throughout the winter. (See image below and also *globular star cluster*.)

Left: M45, the famous Pleiades open star cluster in Taurus. Easily visible with the naked eye throughout the winter, this image shows the deep blue nebulosity that surround the stars. This nebulosity is the remains of the cloud that gave birth to the stars themselves, but unfortunately this is not visible to the vast majority of observers. Photo by the author using Slooh.

Opposition

An object is said to be at opposition when it is directly opposite the Sun in the sky. On that date, it is visible throughout the night as it will rise at sunset, culminate at midnight and set at sunrise. For that reason, this is the best opportunity to observe that object. (See also *conjunction*.)

Perihelion

The point at which an object is closest to the Sun. (See also *aphelion*.)

Perigee

The point at which an object is closest to the Earth. (See also *apogee*.)

Prograde Motion

Prograde motion is when a body appears to move forwards through the constellations from west to east. It's the normal motion of the Sun, Moon, planets and asteroids across the sky. (See also *retrograde motion*.)

Retrograde Motion

Retrograde motion is when a body appears to move *backwards* through the constellations, from east to west. For the inferior planets Mercury and Venus, this happens for a time after greatest eastern elongation (when the planet appears in the evening sky) and before greatest western elongation (when it appears in the pre-dawn sky) as the planet catches up to and then passes the Earth in its orbit. For all the other planets and asteroids, it happens for a time before and after opposition when the Earth catches up to that world and then passes it. This YouTube video from 2009 does a good job of graphically depicting how this happens. (See also *prograde motion*.)

Superior Conjunction

Superior conjunction occurs when either Mercury or Venus are on the opposite side of the Sun from the Earth. For example, if Mercury is at superior conjunction, the Sun would be directly between the Earth and the Mercury. Like *inferior conjunction*, the planet appears very close to the Sun in the sky and is not visible from Earth.

Again, like *inferior conjunction*, Mercury and Venus are the only two planets that can go through inferior conjunction because only these two worlds orbit closer to the Sun than the Earth. (See also *conjunction* and *inferior conjunction*.)

Universal Time

Universal Time is the standard method of notating when an astronomical event takes place. It is based upon Greenwich Mean Time and requires adjustment for other time zones:

Greenwich Mean Time – no change. (Summer Time – add one hour.)

Eastern Time – deduct five hours. (Summer Time – deduct four hours.)

Central Time – deduct six hours. (Summer Time – deduct five hours.)

Mountain Time – deduct seven hours. (Summer Time – deduct six hours.)

Pacific Time – deduct eight hours. (Summer Time – deduct seven hours.)

A useful website that will convert Universal Time to other time zones can be found at the following address: http://www.worldtimeserver.com/convert_time_in_UTC.aspx

Bear in mind that if an event takes place during the day at your location, it may still be visible in the evening or pre-dawn sky. For example, two planets may be at their closest at 1pm local time but because they don't move quickly, they'll still be very close together during the night. The only exception is the Moon – it *does* move relatively quickly, but may still appear fairly close to the object when it next becomes visible. It just won't be as close as it was at the time listed in the book.

Waning

The Moon is said to be "waning" between the full and new Moon. When the Moon is full, the Earth lies between the Moon and the Sun and the lit surface is completely visible to us. When the Moon wanes, the visible, lit portion of the Moon appears to decrease until it is completely invisible at new Moon. A waning Moon is best seen in the pre-dawn sky. (See also *waxing.*)

Waxing

The Moon is said to be "waxing" between the new and full Moon. When the Moon is new, it lies between the Earth and the Sun and the lit surface is not visible to us. When the Moon waxes, the visible, lit portion of the Moon appears to be increasing until it is completely lit at full Moon. A waxing Moon is best seen in the evening sky. (See also *waning.*)

Zenith

The point directly overhead in the sky. (See also *zenith hourly rate.*)

Zenith Hourly Rate

The number of meteors an observer can expect to see each hour at the zenith (directly overhead) on the shower's peak date. It's worth remembering that meteor showers can be somewhat unpredictable and, hence, the maximum zenith hourly rate is only an estimate at best. (See also *zenith.*)

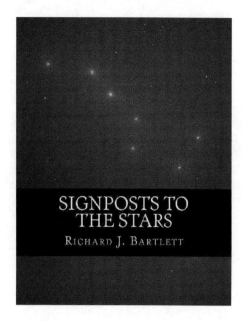

Signposts to the Stars

Aimed at absolute beginners, this book will help you to locate and learn the constellations using the brightest stars of Ursa Major and Orion as signposts.

More than that, the book also details:
*Key astronomical terms and phrases
*The brightest stars and constellations for each season
*The myths and legends of the stars
*Fascinating stars, star clusters, nebulae and galaxies, many of which can be seen with just your eyes or binoculars
*An introduction to the planets, comets and meteor showers

If you've ever stopped and stared at the stars but didn't know where to begin, these signposts will get you started on your journey!

Easy Things to See With a Small Telescope

Specifically written with the beginner in mind, this book highlights over sixty objects easily found and observed in the night sky. Objects such as:
* Stunning multiple stars
* Star clusters
* Nebulae
* And the Andromeda Galaxy!

Each object has its own page which includes a map, a view of the area through your finderscope and a depiction of the object through the eyepiece.

There's also a realistic description of every object based upon the author's own notes written over years of observations.

Additionally, there are useful tips and tricks designed to make your start in astronomy easier and pages to record your observations.

If you're new to astronomy and own a small telescope, this book is an invaluable introduction to the night sky.

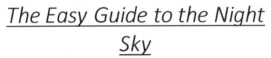

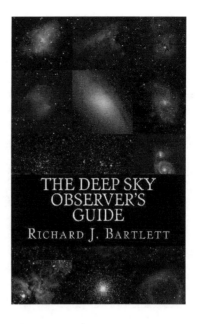

The Easy Guide to the Night Sky

Written for the amateur astronomer who wants to discover more in the night sky, this book explores the constellations and reveals many of the highlights visible with just your eyes or binoculars.

The highlights include:
* The myths and legends associated with the stars
* Bright stars and multiple stars
* Star clusters
* Nebulae
* Galaxies

Each constellation has its own star chart and almost all are accompanied by graphics depicting the highlights and binocular views of the best objects.

Whether you're new to astronomy or are an experienced stargazer simply looking to learn more about the constellations, this book is an invaluable guide to the night sky and the stars to be found there.

The Deep Sky Observer's Guide

The Deep Sky Observer's Guide offers you the night sky at your fingertips. As an amateur astronomer, you want to know what's up tonight and you don't always have the time to plan ahead. The Deep Sky Observer's Guide can solve this problem in a conveniently sized paperback that easily fits in your back pocket. Take it outside and let the guide suggest any one of over 1,300 deep sky objects, all visible with a small telescope and many accessible via binoculars.

* Multiple stars with 2" or more of separation
* Open clusters up to magnitude 9
* Nebulae up to magnitude 10
* Globular clusters up to magnitude 10
* Planetary nebulae up to magnitude 12
* Galaxies up to magnitude 12
* Includes lists of deep sky objects for the entire sky with R.A. and declination for each and accompanying images for many

Whether you use a GoTo or prefer to star hop, no matter where you live in the world and no matter what time of year or night, the Deep Sky Observer's Guide is the indispensable companion for every adventure among the stars.

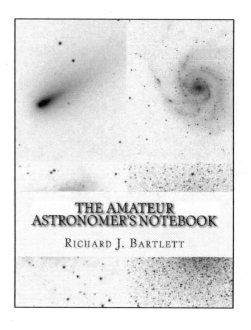

The Wonder of it All

The Amateur Astronomer's Notebook

From our home here on Earth, past the Sun, Moon and planets, this is a journey out to the stars and beyond.

A journey of discovery that shows us the beauty and wonder of the cosmos and our special and unique place within it.

Written by an amateur astronomer with a life-long love of the stars, The Wonder of It All will open your child's eyes to the universe and includes notes for parents to help develop an interest in astronomy.

The Amateur Astronomer's Notebook is the perfect way to log your observations of the Moon, stars, planets and deep sky objects.

With an additional appendix with hundreds of suggested deep sky objects, this 8.5" by 11" notebook allows you to record everything you need for 150 observing sessions under the stars:
*Date
*Time
*Lunar Phase
*Limiting Magnitude
*Transparency
*Seeing
*Equipment
*Eyepieces
*Additional Notes
*Pre-drawn circles to sketch your observations
*Plenty of room to record your notes and impressions

Whether you're an experienced astronomer or just beginning to discover the universe around us, you'll find the notebook to be an invaluable tool and record of your exploration of the cosmos.

72047533R00108

Made in the USA
Lexington, KY
27 November 2017